NEC4 Resolving and Avoiding Disputes

NEC4 Resolving and Avoiding Disputes

Robert Alan Gerrard and Patrick Waterhouse

Published by ICE Publishing, One Great George Street, Westminster, London SW1P 3AA.

Full details of ICE Publishing representatives and distributors can be found at: www.icebookshop.com/bookshop_contact.asp

Other titles by ICE Publishing:

NEC4 Practical Solutions
R A Gerrard and S Kings. ISBN 978-0-7277-61996
Managing Reality. Book Three: Managing the Contract. Third Edition (2017)
B Mitchell and B Trebes. ISBN 978-0-7277-6186-6
NEC3 and NEC4 Compared
R A Gerrard. ISBN 978-0-7277-6201-6

www.icebookshop.com

A catalogue record for this book is available from the British Library

ISBN 978-0-7277-6404-1

© Thomas Telford Limited 2019

ICE Publishing is a division of Thomas Telford Ltd, a wholly-owned subsidiary of the Institution of Civil Engineers (ICE).

All rights, including translation, reserved. Except as permitted by the Copyright, Designs and Patents Act 1988, no part of this publication may be reproduced, stored in a retrieval system or transmitted in any form or by any means, electronic, mechanical, photocopying or otherwise, without the prior written permission of the Publisher, ICE Publishing, One Great George Street, Westminster, London SW1P 3AA.

This book is published on the understanding that the authors are solely responsible for the statements made and opinions expressed in it and that its publication does not necessarily imply that such statements and/or opinions are or reflect the views or opinions of the publisher. While every effort has been made to ensure that the statements made and the opinions expressed in this publication provide a safe and accurate guide, no liability or responsibility can be accepted in this respect by the authors or publisher.

While every reasonable effort has been undertaken by the authors and the publisher to acknowledge copyright on material reproduced, if there has been an oversight please contact the publisher and we will endeavour to correct this upon a reprint.

Cover photo: The Infinity Bridge at Stockton-on-Tees, England, at night.
AC Images/Alamy Stock Photo

Commissioning Editor: Michael Fenton
Production Editor: Madhubanti Bhattacharyya
Marketing Specialist: April Asta Brodie

Typeset by The Manila Typesetting Company
Index created by Pierke Bosschieter
Printed and bound in Great Britain by Latimer Trend, Plymouth

Contents

	Foreword	ix
	About the authors	xi
	Abbreviations	xiii
	Acknowledgements	xv
	Authors' note	xvi

01 An introduction to the NEC4 Engineering and Construction Contract — 1

1. Introduction — 1
2. The objectives of NEC contracts — 1
3. The structure of an ECC — 2
4. People — 3
5. Relationships with third parties — 3
6. Time — 4
7. Quality — 4
8. Payment — 5
9. Change — 5
10. Liabilities — 6
11. Termination — 6
12. Relevant secondary Options — 7
13. Dispute resolution — 7
14. The primary issues in this book — 8

02 Communications — 9

15. Introduction to communications — 9
16. The ECC's requirements for communications — 11
17. Clause 14.3 — 15
18. The importance of communication in information modelling and Clause X10 — 15
19. The importance of communications in disputes — 16
20. Conclusion — 18

03 Early warnings — 21

21. What are early warnings and why do we need them? — 21
22. What should happen once an early warning has been given? — 23
23. The early warning process in practice — 24
24. What happens when the *Project Manager* and the *Contractor* don't comply with the early warning processes? — 25
25. How does the failure to give early warnings affect the Parties' relationship? — 28

04 Programme — 29

26. When is the first programme required? — 29
27. The terminology of time and programmes in the ECC — 30
28. What is an Accepted Programme? — 31
29. What should programmes show? — 31

	30.	What is the role of programmes in disputes?	33
	31.	Keeping the programme up to date	33
	32.	The authority of the *Project Manager* to make changes	34
	33.	Access	34
	34.	Take over	35
	35.	Acceleration	36
	36.	Key Dates	37
05	**Payments**		**39**
	37.	Introduction to payments	39
	38.	Payment provisions in the core clauses	40
	39.	The application and assessment processes	40
	40.	The Price for Work Done to Date	41
	41.	Payment processes in the core clauses	44
	42.	Defined Cost	44
	43.	The Fee	45
	44.	Final assessments	45
	45.	Clause Y(UK)2	48
	46.	The main Option provisions	49
	47.	Main Option A – priced contract with Activity Schedule	49
	48.	Main Option B – priced contract with Bill of Quantities	52
	49.	Assessment of Defined Cost under main Options C, D and E	54
	50.	Defined Cost – dispute resolution	61
	51.	Main Option C – target contract with Activity Schedule	61
	52.	Main Option D – target contract with Bill of Quantities	62
	53.	Main Option E – cost reimbursable contract	62
	54.	Main Option F – management contract	63
	55.	X clauses	64
	56.	Schedules of cost components	64
	57.	Cost finalisation in main Options C, D, E and F	65
	58.	Z clauses	66
06	**Compensation events**		**69**
	59.	Where are compensation events described in the contract?	70
	60.	Additional compensation events in main Options B and D	85
	61.	Additional compensation events in secondary Options	87
	62.	Processes for compensation events	89
	63.	Notification	89
	64.	Quotation	91
	65.	Assessment	92
	66.	The *Project Manager's* assessments	96
	67.	Proposed instructions	97
	68.	Implementation	98
	69.	Time restrictions on the *Project Manager* and extending time	98
	70.	Dealing with compensation events in adjudication and at the *tribunal*	99

07		**Termination**	**101**
	71.	Termination of contracts generally	101
	72.	Termination in the NEC4 ECC	101
	73.	What is being terminated?	101
	74.	Termination for convenience	102
	75.	Starting the termination	102
	76.	The reasons for termination	103
	77.	The *Project Manager's* role in termination	103
	78.	Procedures on termination	104
	79.	Payments due on termination	105
	80.	Where do disputes arise in termination?	106
08		**NEC4 dispute procedures**	**107**
	81.	Common dispute resolution procedures?	107
	82.	What procedures are used in what NEC4 contracts?	108
	83.	Summary	108
09		**The *tribunal* – arbitration or litigation?**	**111**
	84.	Resolving disputes through arbitration or litigation	111
	85.	Avoiding disputes through arbitration or litigation	114
10		**The *Senior Representatives***	**117**
11		**Adjudication in Option W1**	**121**
	86.	Avoiding disputes in Option W1 adjudication	121
	87.	Resolving disputes in Option W1 adjudication	121
	88.	The *Adjudicator*	121
	89.	The adjudication	122
	90.	How have the courts responded to adjudication?	124
	91.	Summary	125
12		**Adjudication in Option W2**	**127**
	92.	Avoiding disputes in Option W2 adjudication	127
	93.	Resolving disputes in Option W2 adjudication	127
	94.	The *Adjudicator*	128
	95.	The adjudication	129
	96.	How have the courts responded to adjudication?	131
	97.	Summary	131
13		**The Dispute Avoidance Board in Option W3**	**133**
14		**The NEC4 Dispute Resolution Service Contract (DRSC)**	**139**
	98.	Avoiding disputes using the DRSC	139
	99.	Resolving disputes using the DRSC	139
	100.	Resolving disputes – adjudication	141
	101.	Resolving disputes – Dispute Avoidance Board	142
	102.	Summary	142

		Secondary Options	**145**
15	103.	Avoiding disputes using secondary Options	145
	104.	Option X1 – price adjustment for inflation (used only with Options A, B, C and D)	145
	105.	Option X2 – changes in the law	146
	106.	Option X3 – multiple currencies (used only with Options A and B)	146
	107.	Option X5 – sectional Completion	146
	108.	Option X6 – bonus for early Completion	146
	109.	Option X7 – delay damages	147
	110.	Option X10 – information modelling	147
	111.	Option X12 – multiparty collaboration (not used with Option X20)	148
	112.	Option X14 – advanced payment to the *Contractor*	148
	113.	Option X15 – the *Contractor's* design	149
	114.	Option X17 – low performance damages	149
	115.	Option X18 – limitation of liability	149
	116.	Option X20 – Key Performance Indicators (not used with Option X12)	150
	117.	Option X21 – whole life cost	150
	118.	Option X22 – early *Contractor* involvement (used only with Options C and E)	150
	119.	Option Y(UK)1 – project bank account	151
	120.	Option Y(UK)2 – Housing Grants, Construction and Regeneration Act 1996	151
	121.	Option Z – additional conditions of contract	151
	122.	Resolving disputes using secondary Options	151
	123.	Summary	152
	Index		**153**

Foreword

There is a real possibility that this book will do more good for standard form contracts than any other. Its theme is dispute avoidance; its approach, its tone, and its style is building-contractor-friendly. Forms of Contract are written by lawyers for building Buildings. Those two camps aren't always able to fathom each other. In the end the builder just gets on with building. Whether it is this or that contract document is all very well but spending time trying to operate contract clauses in the 'muck and bullets' of building something is, er, hmm, not a priority. It's only when a dispute breaks ground that building folk start rummaging the small print . . . dozens and dozens of pages in dozens and dozens of contract documents. It might even be that someone reaches for a textbook. Ah, then comes the lawyer stuff.

In 1991, came the '*New Engineering Contract*' ('NEC'). The NEC4 Edition was 2017. In all these years, the NEC folk have said that the intention is to operate NEC so as to minimise disputes. Marred by-the-way, say the two authors of this book, by '*Z clauses (having) gained a certain notoriety amongst NEC users, as they frequently used to skew the contractor's risk allocation between the Parties.*' So, the book has, at its heart, how to understand NEC, how to enjoy using it as a hand-book and yes, how to avoid, or at least reduce disputes. Dispute avoidance is, at last, gaining ground.

Tony Bingham
Arbitrator, Adjudicator,
Barrister at Law

About the authors

Patrick Waterhouse CEng FICE FCIArb FCInstCES is a construction adjudicator and contract consultant. He is a recognised expert in the NEC forms of contract. He has adjudicated numerous disputes under NEC contracts and has acted as party adviser in many more. He delivers training in all forms of contract and is an examiner for the ICE's law and contract management examinations.

Robert Alan Gerrard BSc (Hons) FRICS FCInstCES has been the NEC Users' Group Secretary since 2005 and was part of the NEC4 drafting team. He is an NEC consultant and has run numerous NEC training courses, working with NEC contracts since 1996.

Abbreviations

ALC	NEC4 Alliance Contract (June 2018)
DRSC	NEC4 Dispute Resolution Service Contract (June 2017) (with amendments January 2019)
ECC	NEC4 Engineering and Construction Contract (June 2017) (with amendments January 2019)
ECSC	NEC4 Engineering and Construction Short Contract (June 2017) (with amendments January 2019)
SCC	Schedule of Cost Components
SSC	NEC4 Supply Short Contract (June 2017) (with amendments January 2019)

Acknowledgements

In writing a book on this subject we have been keen to reflect the real work of NEC contracts. A paucity of case law in comparison to other standard forms could leave you to conclude that NEC contracts do not generate many disputes. Those working in the field of dispute resolution know otherwise. For reasons which we explain in this book, there is generally a reluctance of disputing parties to move to litigation after the resource-intensive process of adjudication which is a condition precedent of litigation in NEC contracts. But we also wanted to reflect that NEC contracts can and do generate collaborative working and many success stories.

This book demonstrates that the methods needed to avoid disputes are also helpful in managing the dispute as and when it happens. We therefore hope that readers will benefit from our work regardless of the aspect of this book's title they come from.

We are both seasoned trainers of all things NEC. We are both accredited mediators for commercial disputes. These experiences have convinced us that parties' commercial objectives are best served by ensuring that the various participants are knowledgeable about the contract and that they maintain a constructive dialogue throughout.

Patrick's thousands of hours adjudicating construction contracts disputes, many of which have been under an NEC form, have shown him the consequences of parties not doing what they have committed to. Unlike some other standard forms of contract, the NEC forms contain a sanction for each party's failure to comply with its provisions. From an adjudicator's point of view that makes the assessment relatively easy in certain circumstances, where a party is in breach. But that can lead to at least one party concluding that its 'punishment' for failing to comply has been disproportionate.

Of course it is preferable if adjudication, time bars, damages etc. can be avoided. We hope that this book will sit alongside other texts, guidance notes, user guides etc. for practitioners. We also hope that this book will be used by those working in dispute resolution, both legally qualified and not. But our aim is also to support ordinary users of NEC contracts and to hopefully minimise their contact with those in dispute resolution.

We have benefited from our contact over the years with many people in the industry working with NEC contracts. Our thanks go to them for our opportunity to learn from their experiences and we have endeavoured to reflect the real world in our text.

Finally we thank our respective families for their patience and understanding during our writing of this book.

Robert Gerrard
Patrick Waterhouse
March 2019

Authors' note

In this book, since many defined terms are *italicised*, underlining has been used for emphasis.

NEC4 Resolving and Avoiding Disputes

Gerrard and Waterhouse
ISBN 978-0-7277-6404-1
https://doi.org/10.1680/necrad.64041.001
ICE Publishing: All rights reserved

Chapter 1
An introduction to the NEC4 Engineering and Construction Contract

1. Introduction

The NEC4 Engineering and Construction Contract (ECC) is the fourth generation of a contract conceived in the mid 1980s and first published in 1991. Initially, there was the 'New Engineering Contract', from which the NEC name was derived. The principles of that initial contract were then rolled out to a series of related contracts for the industry when NEC2 was born in 1995. Much has been written on the provenance of the ECC so we will not provide a detailed history lesson here. But the New Engineering Contract (as it then was) introduced adjudication to the UK construction industry in 1993, a full five years before the Housing Grants, Construction and Regeneration Act 1996, as amended,[i] created a statutory right to adjudication for contracting parties in England, Wales and Scotland.

The objectives of all NEC contracts are to help the Parties to avoid disputes by focusing on the stimulation of behaviours that make a positive impact on projects. But in commercial life disputes can and do occur. Contract documents and the Parties must anticipate the possibility of disputes and map out a way of resolving them.

Dispute avoidance makes commercial common sense. Disputes take time, people and money to resolve. If they can be avoided in the first place, then all parties are presumably better off. Disputes can be avoided; with any form of contract, following its requirements should maximise the probability of doing so. NEC contracts, with their focus on collaboration, perhaps increase this probability but cannot guarantee that the Parties' relationship will remain harmonious. Commercial life brings its own pressures and the Parties' interests may not always be best served by following the contract. Where such incidents develop into disputes, the contract must contain provisions to deal with them.

This book has two subjects; avoiding disputes and resolving them. The two are inevitably closely linked. Correctly managing the contract will always minimise the chances of experiencing a dispute. It is a recurring theme of this book that the correct management of the contract will usually also prepare a Party well for dispute resolution. The ability to demonstrate one's own case in legal proceedings is as important, if not more so, than the case itself. Giving the requisite early warnings, issuing programmes on time etc. are ways of (hopefully) avoiding disputes but also laying the groundwork for later demonstrating compliance to an adjudicator, arbitrator or judge.

2. The objectives of NEC contracts

The three objectives of all NEC contracts are as follows.[ii]

- *They stimulate good management* of the relationship between the two Parties to the contract and, hence, of the work involved in the contract.

- *They can be used in a wide variety of commercial situations*, for a wide variety of types of work and in any location.
- *They are clear, simple and written in plain English*, using language and a structure that is straightforward and easily understood.

These objectives would seem to assist the Parties in forming and managing a positive relationship.

3. The structure of an ECC

The stimulation of good management in the ECC exists primarily as a disincentive to bad management. Not doing what the contract requires tends to cost each Party money, put simply. A series of sanctions against the Parties acts to deter the wrong sort of behaviour. Concepts such as Disallowed Cost or the deemed acceptance of compensation event communications make the defaulting Party responsible for its own actions.

The flexibility of use brings its own challenges to Parties at the formation stage. Getting the structure of the contract right for a project will always assist a positive relationship. Conversely, getting it wrong will hamper the Parties' efforts. The flexibility works in two respects; the type of work involved and the nature of the commercial relationship between the Parties. We have encountered the use of the ECC in a wide range of engineering activities. There is the usual construction of roads, railways, airports, power stations, hospitals, schools, flood defences etc. But we have also seen it deployed successfully in the transport of spent nuclear fuels and the provision of a logistics network supporting the construction of a power station. The ECC's language and its provisions are sufficiently flexible to support any form of engineering or construction activity.

All NEC contracts contain an opening statement in one of the cover pages explaining what the particular contract is designed for. We are told by the publishers that the ECC 'should be used for the appointment of a contractor for engineering and construction work, including any level of design responsibility.' Potential users of the contracts should follow this advice closely. The NEC contracts have been developed to meet particular specifications; using the wrong contract could set the Parties off in their new relationship on the wrong path.

The flexibility of the commercial relationship requires the Parties, but more usually the *Client*, to make decisions on the clauses to include in a contract. The main Option is key; do we want a cost reimbursement contract or a fixed price contract? Should we use a target contract to share the payment risk? All these questions require careful analysis. The selection of the correct main Option (if indeed there is such a thing) is a major decision for the *Client* to make. If the Scope is not sufficiently defined, the tenderers will have to make decisions on the pricing and management of risk. Assumptions will be made; if the appointed *Contractor* starts to lose money on the project, it may amend its behaviour to reduce or recover its losses. This, in turn, may lead to disputes. The selection of the main Option must match the specific features of the project.

A similar approach to the selection of secondary Option X, Y and Z clauses must also be adopted. They have been written so that each clause, or almost any combination[iii] of clauses, can be included in an ECC contract. Before including an X clause in a draft contract, the *Client* should consider its purpose and the incentives or behaviour that it will create.

Perhaps the most contentious decisions made by potential clients at the initial stages of procurement are the Z clauses. Included to amend a standard NEC form, Z clauses are designed to ensure that a contract meets the needs of the Parties. Many Z clauses provide a sensible and balanced addition or amendment to the contract to reflect specific needs, for example in respect of the *Client's* funding (say, access for funding monitors) or security provisions in military projects.

But Z clauses have also gained a certain notoriety among NEC users as they are frequently used to skew the contract's risk allocation between the Parties. Tricks of the trade involve reducing the *Contractor's* response times (for example to notify a compensation event in Clause 61.3) or increasing the *Project Manager's* response times (for example in accepting a programme in Clause 31.3). Where clients amend contracts in this manner they delude themselves if they believe that it will improve the chances of the project being successful. Each decision of this nature drives slightly different behaviours from the participants than are initially proposed by the contract's publishers. At the time of publishing this book, NEC contracts have been in use for 26 years and the various time provisions have been well proven.

When clients make decisions on any aspect of the contract's structure, they should consider whether each decision makes a positive influence on the Parties' ultimate relationship and act accordingly.

4. People

We see repeatedly that one of the major determinants of success for a construction project is people. The ECC provides the *Client* with the ability to require the deployment of people identified by the *Contractor* in its tender. *Key persons* identified in the Contract Data must be used by the *Contractor* or, where a person isn't available, the replacement must be accepted by the *Project Manager*.

Many tender competitions for construction projects involve a balanced approach, analysing both price and quality. The quality submissions can focus on people; their qualifications, experience and so on. By identifying the *key persons* at the tender stage, the *Client* and the *Contractor* should have some certainty as to the management team.

While good people are a key determinant of success, unfortunately the reverse of that coin is that we tend to see difficulties when people issues are not managed well. The *Project Manager* and the *Contractor* should work together closely to operate these properly. The *Contractor* should make an early notification when a key person is to be replaced. The *Project Manager* should not use the acceptance provisions unfairly; an example of the importance of Clause 10.2.

5. Relationships with third parties

No construction project exists in a vacuum and its relationships with the outside world are important. The ECC defines third parties as 'Others'[iv] and places obligations on the *Contractor* to manage interactions with Others where relevant. The definition of 'Others' also includes other suppliers, designers or constructors engaged by the *Client* on the same project.

Managing relationships with third parties can mean many things, from being a good neighbour (e.g. minimising dust, fumes, dirt on roads etc.) to major issues like structural damage to adjacent property caused by construction methods.

In addition to the provisions for liaising with third parties, the ECC now includes (for the first time in NEC4) undertakings to the *Client* or Others (secondary Option X8), whereby Subcontractors or the *Contractor* make parallel commitments to the *Client* or Others, respectively. In the UK, and other legal jurisdictions, these are often called collateral warranties.

The ECC, as with many other NEC contracts, includes clauses allowing for multiparty collaboration. This usually incentivises collaboration between organisations working on the same project. The clauses, in secondary Option X12, have been renamed from 'partnering' to 'multiparty collaboration'.

All these contractual provisions are designed to provide the basis of the effective management of relationships with third parties. But they can be a very busy area for those of us working in disputes. Once third parties are involved in a problem, the two Parties to the ECC lose some of their ability to resolve issues in a collaborative fashion. The demands of a third party may force one or both Parties into a position that otherwise might not have happened. Understanding the obligations to third parties and managing them properly are key to avoiding disputes in this area.

6. Time

All NEC contracts contain detailed provisions on the management of time. Differentiating, for example, the Contract Date, the *starting date* and the *access date(s)* creates an environment where the Parties know what to expect of one another and by when. A similar logic applies to the distinction between Completion and take over. Where these dates aren't clear, for example in other standard contract forms, it leads to disagreement and dispute. The inclusion of Key Dates assists the *Client* with the co-ordination of several suppliers on the same project, using the unliquidated damages provisions to keep the *Client* unharmed in the event of a delay by one or more of those suppliers.

The programme provisions are extensive; while it is the responsibility of the *Contractor* to prepare and maintain the programme (it is a forecast of the *Contractor's* work), the programme should be capable of being relied on by the *Client*, *Project Manager* and *Supervisor*. The regular updates required of the *Contractor* ensure that the programme is kept up to date and isn't just a glossy reminder of the heady days of the contract being formed.

Programmes, time and completion are all regular components of disputes in construction projects; therefore, the likelihood of disputes can be reduced by using carefully drafted clauses like those in Section 3 of the ECC. Acceleration by the *Contractor* may be requested by the *Project Manager* or the *Contractor*, but the changed obligation cannot be imposed by either. Again, if one Party is unable to amend the *Contractor's* obligations unilaterally, this reduces the likelihood of dispute.

The time provisions link to several other areas of the contract, most noticeably compensation events, where the time and cost implications of change and risk events must be considered together.

7. Quality

The quality provisions have been robustly improved in the ECC as compared with earlier editions of the contract. Recognising the development in management systems and their certification, the new contract requires the *Contractor* to operate a quality management system, to have a quality policy statement and to prepare a quality plan. The plan must be kept up to date.

None of these terms is defined or explained; the authors of the contract anticipate that the Parties will do this for themselves, either in the Scope or as Z clauses. Quality is a subjective process; therefore, those writing items for the Scope need to prepare the words accordingly, relying on objective tests where possible. As with many areas of NEC contracts, the Parties must get 'hands on' with creating contract documents to ensure that the contract matches the needs of the Parties and of the project. Where such drafting is of a high quality, each Party will know and understand what the other wants and should be better placed to provide it and hence the potential for disagreement will be reduced. Of course, quality provisions are not an end in themselves but, if undertaken correctly, should increase the likelihood of the project being completed successfully.

8. Payment

Payment. The root of all evil in the industry. Or is it? Cash flow, someone once said, is the lifeblood of the construction industry. It is probably the lifeblood of many other industries too, but our industry tends to make life very difficult through the lack of cash at times. Many legal jurisdictions have legislated to intervene, in the freedom that otherwise exists, that parties to a contract have to agree terms. Even where such law does not exist, local markets tend to develop their norms. These norms often create long-term expectations and behaviours, many of which are not respected by NEC.

The ECC contains six main Options, of which only one should be used at a time. We have seen contracts where someone has decided to use more than one Option. Often this is split, with one part of the project being price-based and the other cost reimbursable. The Parties then have the chore of forensic allocation of Defined Cost to either pot, drawing on a significant amount of professional resource, which does not always see eye to eye on things. As with other uses of the contract that weren't envisaged by the original drafters, this rarely work well.

Each main Option allocates the payment quantum risk between the Parties in different ways. The contract contains extensive provisions for transparency of cost information in the four main Options where the *Client* shares the risk of the final outturn cost with the *Contractor*. Open co-operation of the Parties assists in building confidence and avoiding disagreement. Of course, where disputes do arise, it is helpful that the *Contractor* is obliged to have kept records and to produce them where necessary. Where the *Contractor* has done so then, of course, the *Client*, the *Project Manager* and Others will need to produce their records accordingly. There is nothing new about these necessities but NEC contracts are rare in their express requirements for such records to be maintained.

The payment provisions provide clear allocation of responsibility and time periods for the Parties and *Project Manager* to react. The concept of Defined Cost applies to all main Options and Disallowed Cost to four of the main Options. Final assessments were introduced to the ECC in NEC4, allowing the Parties to draw an effective line under selected periods of the project (for main Options C, D, E and F) or for the entire project (all main Options). This allows the list of items for debate to be reduced and, if operated properly, will enhance the relationship between the Parties.

9. Change

A book about disputes will never stray far from the subject of change and this one is no exception. Parties' expectations about the other's obligations will always differ; the role of a well-drafted contract is to set out the obligations for both Parties. Inevitably, the subject will boil down to one of money and a primary component of that is time. The ECC's compensation events deal with time and money together. The provisions are lengthy, but necessary. By delineating the process into four stages, the drafters have created a system that works well with the requisite notices having been notified by the *Project Manager* and the *Contractor*. Unfortunately, where these two have not done what the contract requires of them, the opposite will apply.

Given the obvious link between change and disputes, Parties must resource the compensation events processes properly and pay due regard to honouring the timescales and the noticing obligations. These are tasks that require skilled people who have been trained in what to do. Too often we see disputes that were initiated by people who were ill-suited to their tasks and who were subsequently backed up by colleagues who knew no better.

Adjudicators, arbitrators and judges will all look for evidence of what a Party has done, or not done. Those who have stayed close to the requirements of a contract are always going to fare better than those who have not. Once again, adhering to the contract will assist with avoiding disputes and will assist in resolving them if they occur.

10. Liabilities

The liabilities of the Parties are set out in Clauses 80 and 81, albeit the remainder of the *conditions of contract* also attach liability to the Parties. Significantly rewritten in the NEC4 update, this section of the contract provides clarity, particularly in the *Contractor's* liabilities to the *Client* and to Others. Many of the *Contractor's* liabilities will rely on the *Contractor's* insurances to be met; therefore, the *Client*, through the *Project Manager*, must ensure that those insurances are in place and remain in place until the dates stated in the Contract Data. Additional *Client* liabilities can be stated in the Contract Data; if this possibility is taken, the drafting must be clear and consistent with the ECC's other provisions. Disputes involving liabilities such as those in Section 8 of the ECC can generate very high values and will typically involve insurers.

The key to avoiding disputes here lies in ensuring that the Party (or Parties) who should provide insurance does so and that any contractual and extra-contractual notifications are issued on time.

11. Termination

Termination is a subject that thankfully appears rarely in most professionals' lives but, when it does arrive, they are invariably ill-prepared for it. NEC contracts contain provisions to terminate the *Contractor's* future involvement. This is distinct from terminating the contract itself, which, in most legal jurisdictions, remains a possibility for reasons independent of the contract's own terms.

Termination is a fraught process, even in the, very rare, scenario where the need to terminate is agreed. Parties will disagree on payment and on timescales and the seemingly straightforward clauses in Section 9 of the ECC will very quickly seem hard to decipher. While both of the authors are keen to keep lawyers at bay in the day-to-day running of contracts, termination is not a subject for amateurs and legal input should be sought at an early stage.

Those managing contracts should remain alert for the signs of insolvency, one of the major events leading to termination. Contract managers should also be involved if termination is being considered for reasons of breach by the *Contractor*. Several of the reasons for termination rest on the conduct or failure of the *Contractor*, the *Client* and the *Project Manager*. Each should remain aware of the potential implications of their conduct where termination is concerned.

12. Relevant secondary Options

The ECC contains 22 secondary Options denoted 'X' and three clauses denoted 'Y'. We explain these in more detail later in the book but as an introduction it is necessary to emphasise the importance of using these clauses in the right way and of providing the necessary additional information elsewhere in the contract.

For example, the *Client* may want a performance bond (Clause X13), an ultimate holding company guarantee (Clause X4), delay damages (Clause X7) and low performance damages (Clause X17). The structure and drafting of the ECC allows this combination and it should create an enforceable contract. But what message might this 'package' send to tenderers? How might this affect the behaviour of the *Contractor*? This isn't to say that such combinations are to be avoided, but they need to be used appropriately.

By way of example, look at Clause X13 (Performance bond). The words clearly require (i) the amount to be stated in the Contract Data and (ii) the form to be set out in the Scope. These details need to be correct, otherwise it is likely that differences in interpretation will occur and the bond may not be put in place. Among many other implications, this may justify termination. All of this could be avoided with careful drafting in the Contract Data and Scope. Similar administrative requirements are also associated with other secondary Options.

13. Dispute resolution

While the ECC is well suited to assisting the Parties in avoiding disputes, its provisions for resolving them are also aimed at the Parties co-operating to do so. That co-operation cannot be taken for granted, as we see in many disputes; therefore, the adjudication provisions can be concluded, even with the reluctant involvement, or even no involvement, of one of the Parties.

The role of the *Senior Representatives* has been introduced in NEC4, a technique that has worked in other commercial contracts for many years.[v] The mere identification of these people won't, on its own, provide a useful input unless those people genuinely are senior and are involved in the process.

Adjudication remains the mainstay of the ECC dispute resolution provisions in Options W1 and W2. New to NEC4 is the Dispute Avoidance Board in Option W3, a group of people appointed at the start of a project, regardless of the dispute status of the project. The members of the board visit the Site regularly and remain in touch with progress. If and when a dispute arises, the board makes a recommendation for its resolution. The success of a board depends on the quality of the people appointed to it and the methodology used in its work.

Reference of a dispute to arbitration or to the courts can only be made when the dispute has been decided on by the *Adjudicator* or where the Dispute Avoidance Board has made a recommendation.

14. The primary issues in this book

With each subject addressed in this book, two predominant themes appear repeatedly.

1. Where both Parties follow the procedures and timetables in the contract, the chances of experiencing a dispute are significantly reduced.
2. Where a dispute does occur, a Party that has followed the procedures and timetables in the contract will be better placed to achieve a satisfactory resolution to the dispute.

We return to these points in many of the chapters that follow.

NOTES
 i HMG (Her Majesty's Government) (1996) Housing Grants, Construction and Regeneration Act 1996. The Stationery Office, London, UK. Amended by HMG (2009) Local Democracy, Economic Development and Construction Act 2009. The Stationery Office, London, UK.
 ii NEC (2014) Why NEC? https://www.neccontract.com/About-NEC/Why-NEC (accessed 27/05/2018).
 iii The one exception is that secondary Options X12 and X20 should not be used together; both include Key Performance Indicators and to include both would provide duplication.
 iv 'Others' are people or organisations who are not the *Client*, the *Project Manager*, the *Supervisor*, the *Adjudicator* or a member of the Dispute Avoidance Board, the *Contractor* or any employee, Subcontractor or supplier of the *Contractor* (Clause 11.2(12)).
 v See, for example, contracts under the UK Government's Private Finance Initiative.

Chapter 2
Communications

15. Introduction to communications

Communications are at the heart of all business transactions; the construction industry is no exception to this. All NEC contracts have communications requirements expressly set out with the aim that communication is undertaken efficiently and clearly. Unsurprisingly, communications or, more often the lack of them, play a significant role in contract disputes.

Many commercial contracts still use language suited to postal communications between the Parties and Others. In an era where email and other forms of electronic communication are now the norm, reliance on such drafting is inadvisable, given the likely non-compliance from day one.

For routine matters, the term 'in writing' doesn't appear in the ECC.[i] Instead, the contract requires communications to be in a form that can be read, copied and recorded. The language refers to the functionality of the communication, not the format of its transmission. The focus on functionality continues throughout all NEC4 documents and the ECC is no exception.

We describe the various processes in the contract in later chapters, for example the assessment of payments and compensation events, so we will not cover communications for those processes in this chapter. But it is important for users of the contract to remain aware that all required communications (as opposed to those that are merely desirable) must be provided within a defined timescale. Some clauses contain express timescales, for example the *Project Manager's* notification of the acceptance (or not) of a programme under Clause 31.3 must be communicated within two weeks. Where such a time period has not been specified in the *conditions of contract*, the *period for reply* will establish the time for a response. So the contract defines not only the functionality of the communication, but its speed too.

Disputes in all forms of contract tend to reflect the Parties' actions and inactions. Many NEC processes rely on the protagonists' timely actions. It is common to see allegations in all forms of dispute resolution that the *Project Manager*, *Supervisor* and *Contractor* have not communicated correctly. The NEC's provisions enable the correct provision of communications to be proved or disproved.

Parties to disputes are often left to rely on less-than-perfect evidence to demonstrate that a communication was issued. Notwithstanding the requirements of Clause 13.1, many participants omit important communications and are left to seek out minutes of meetings, drawing transmissions and the like that were never intended to act as communications. Adjudicators are well used to seeing such documents dressed up by party representatives; typically, this is a sign that the correct form of communication was not issued.

NEC contracts separate notifications from other communications required by the contract. The term 'notification' is not defined, and the users of the contract are left to interpret the term themselves. While this may sound uncertain in its design, its application tends to work well. Many clauses in the contract require someone to notify someone else of a fact or a decision. Where the word 'notify' or 'notification' is used, the contract requires it to be sent separately from other communications. This reduces the chance that an important communication will be overlooked because it is low down in an email chain containing other communications. It also prevents the cynical and deliberate 'burying' of some communications in this manner.

The construction industry's methods of communication have changed considerably since the advent of ubiquitous email technology in the 1990s. While prior use of telex and facsimile technology had made their own impacts, email and internet technology have enabled the industry to take massive strides in the last 30 years. The use of online project platforms for sharing and communicating information have enabled faster dissemination of information, online editing and sharing and, crucially for those of us working in disputes, an audit trail for all of this. Information is transmitted in seconds, not days, and modern contracts need to accommodate such speeds. NEC contracts tend to work on the expectation that communications will be electronic, either through email or an online system. But their clauses remain effective for those who wish to rely on paper communications sent through a postal system. The ECC brought in provisions for the Parties to be obliged to use a communication system that is required in the Scope.

Where the use of a communication system has not been stated in the Scope, communications are effective when received. For those relying on paper communications, consideration must surely be given to obtaining a signature on delivery of important communications. It is a common feature of disputes that Party A claims that Party B did not send a communication. Proving its despatch is usually easier than proving its receipt. For these reasons, and many more, it is important that Parties to an ECC contract (and indeed any other commercial agreement) routinely acknowledge receipt of all communications. In this regard, the provisions of Clauses 10.1 and 10.2 are helpful.

Communication systems have much to be commended in the ease of communication and the efficiency that they bring to projects. Use of common databases, repositories etc. all help the Parties to work together and collaborate. From a disputes viewpoint they provide a verifiable record of who sent what, to whom and when. England has already seen its first litigation[ii] concerning such systems and the lessons of that case are worthy of consideration by contract parties the world over. The case concerned an engineering consultant, which hosted the data for a project on its own servers. Following a disagreement over payments, the consultant denied its counterpart access to the data. It seems that the second party had not considered where its data was being held and how it could guarantee access to it. The increasing reliance of the industry on communications technology in its wider sense will lead to further disputes as the Parties discover that their initial expectations for communication have not been matched by their counterpart's actions.

It is an old urban tale that the three most important things needed to succeed in a legal claim are 'records, records and records'. Whether these words were ever actually muttered by a judge or anyone else has never been demonstrated. But they remain as true today as they were 50 years ago when the clear majority of contract communications were made on paper. Storage and retrieval technology should make us more efficient at what we do. Compare, for example, searching lever arch files full of letters for the one document you desperately need with the modern-day equivalent of searching a database in seconds on your PC, tablet or smartphone. However, the use of modern technology has made communicating far easier and it does tend to be less strategic than it might have been in the days of paper. So, generally we have far too much data to mine to find the information that we require. In this regard, the words 'data' and 'information' have very different meanings.

Typically, many of the records, records and records that we are trying to find in disputes are communications, notifications that avoid the operation of a time bar, or early warnings that protect the *Contractor's* full entitlement in compensation events. These are the bread and butter of the people battling contractual disputes. The ECC has the machinery to make these things easy for the participants; it is up to them to use that machinery. Those who disregard the requirements are likely to do so at their own peril; that is certainly the experience of both authors. The aim of this chapter is to identify how communications should be provided and what will be of value when a dispute is being contested in adjudication, arbitration or litigation.

16. The ECC's requirements for communications

The importance of communication is reflected in many of the ECC's clauses, but the starting point is Clause 13.

Before looking at Clause 13, it is perhaps helpful to understand what Clause 11 says about notation. The standard use of capitalised words for defined terms (Accepted Programme, Contract Date, Defect etc.) is stated here. Unique to NEC contracts is the use of italicised terms that are 'identified' in the Contract Data (e.g. *tribunal, defect correction period, starting date*). When communicating with others involved in the contract and making records, it is advisable to adhere to these notation principles to ensure future accurate interpretation. NEC documents occasionally present their users with interpretation challenges (e.g. *completion date* and Completion Date, or Equipment and equipment). In both examples, similar-looking words have slightly different meanings, so careful use of the notation is vital.

Clause 13 is entitled 'Communications' and so is naturally the place that users of the contract refer to when considering if, when and how a communication should have taken place. All three of these considerations are important for both avoiding disputes and, should one arise, resolving a dispute.

Clause 13.1 requires communications to be given in a form that can be 'read, copied and recorded'. Many contracts demand formats such as 'in writing' or 'first-class post'. Except for Clause 12.3, the ECC does none of these things and instead focuses on the functionality of communications. For things to be understood, they need to be capable of being read. For them to be disseminated, they need to be copied. And perhaps most crucially to dispute resolution, they need to be recorded so that they may be retrieved later. These rules apply equally to paper-based communication and electronic means of communication.

Clause 13.2 is new to ECC and reflects the increasing use of electronic communication systems. Many such systems are available, and a small number are bespoke to NEC contracts. Such systems usually increase the efficiency of the project team, but users must think carefully about the full implications of using them. The ECC does not explain what it means by a communication system, but it is assumed that the reference is to the type of online systems that allow the Parties to exchange and store documents, correspondence and the like.

It is worth examining Clause 13.2 in detail as its implication, whether deliberate or not by the drafters, is significant.

> **Clause 13.2**
>
> If the Scope specifies the use of a communication system, a communication has effect when it is communicated through the communication system specified in the Scope.
>
> If the Scope does not specify a communication system, a communication has effect when it is received at the last address notified by the recipient for receiving communications or, if none is notified, at the address of the recipient stated in the Contract Data.

The implication is that, should a communication system be specified in the Scope, it is the only form of communication that is recognised by the contract. The full efficiency of such a system will never be achieved if the various protagonists don't use it. Communications provided by a combination of means (e.g. online system, paper, emails) will always prove difficult to search and verify. By contrast, if everything has been communicated using the communication system, such activities will be greatly simplified.

Whether the drafters of the ECC intended to make communication systems exclusive is unclear, but the Parties should ensure that they differentiate between communications that the 'contract requires' (Clause 13.1) and 'a change to the contract' (Clause 12.3), which must be in writing.

Where a communication system hasn't been specified in the Scope, Clause 13.2 refers to 'the last address' notified by the recipient. This wording is sufficiently broad to imply either a physical address or an email address.

Parties relying on communication systems should consider where the data is stored and how to access it. We have mentioned the *Trant v Mott MacDonald* case elsewhere, which shows how important these issues are. One of the authors was an adjudicator in a dispute where the *Client* blocked the *Contractor's* access to the online document repository, thus hampering the ability of the *Contractor* to demonstrate its case. The contract was silent on the period to retain documents. In reality this period needs to be specified in the Scope if such systems are to be relied on and each participant should have the ability to archive information from the system at a time of its choosing. The adjudication in question therefore started difficultly as the Parties grappled with this problem.

The second part of Clause 13.2 relates to communications in the situation where a communication system has not been specified. Where you want to use the physical transmission of paper communications, proof of delivery is essential.

Frequently in disputes, the very existence of important communications is questioned by the Parties. It is incumbent on the Parties to demonstrate that communications have been transmitted and received correctly. This needs to be thought about at the time as it is usually too late to influence such things when a dispute occurs.

Clause 13.3 provides a default period for the response to communications where such a response is <u>required</u> (emphasis added). Failures to respond timeously to communications, or at all, feature heavily in disputes. A response that says 'I've got your email, I'll respond in due course' is not a response at all. Most actions in the ECC are time-limited. Some, for example compensation events and payments, are limited to bespoke periods stated in those clauses. But where a response is required, and a bespoke period isn't specified, then the *period for reply* will apply to that response. While many disputes feature allegations of failed or late attempts at communications, these can be avoided if communications are managed properly. Inevitably your chance of avoiding a dispute is increased if you (i) comply with the contract and (ii) combine this with some courteous, personal communications. While the latter may not be recognised by the contract, they cannot be divorced from your more formal communications.

Clause 13.4 contains one of the first references in the book to somebody not accepting something. The ECC doesn't refer to the rejection of a communication, but to the recipient choosing not to accept it. This clause requires the *Project Manager* to a reply to a communication 'submitted or resubmitted' by the *Contractor*. Remember that the *Project Manager* must act as stated in the contract.[iii] Where the *Project Manager* does not accept the communication, the response needs to explain why and in sufficient detail. Where such explanations and details are not provided, this leads to confusion and possibly dispute. The *Contractor* will be perplexed if, for example, a perfectly competent design submission is not accepted and no explanation is given. In this scenario, what could the *Project Manager* expect the *Contractor* to do in reworking the submission? Such a lack of clarity will inevitably lead to delay and abortive expenditure. However, the *Project Manager* can withhold acceptance where more information is needed. Getting the balance right in this scenario involves some good old-fashioned verbal communication and collaborative working to minimise any lost time.

Clause 13.5 allows the *Contractor* and the *Project Manager* to agree an extension to the period for reply to a communication. Anyone asking for such an extension should do so as early as possible. Asking for it on, say, day 20 of a 3 week period is likely to be less successful than on day 7.

Clause 13.7 aims to make important communications stand out from other day-to-day exchanges. The ECC does not define what it means by 'notification' or 'certificate'; therefore, there is potentially an interpretation point here. But these terms are used widely in the contract and we must assume that the drafters intended a communication that must be notified (e.g. early warning, Clause 15.1) to be dealt with differently from something that must be submitted (e.g. Equipment design, Clause 23.1). Where a communication involves a notification or a certificate, it should be sent separately from anything else. For example, a *Project Manager* may wish to notify five early warnings. These should be sent as five separate messages. While this may seem excessively bureaucratic, it does allow the *Contractor* to disseminate five messages internally, possibly with five different recipients, and then close each issue separately. Compare that with five messages on a single communication, where in the past we would likely demarcate each and then photocopy the communication four times to put this in the relevant files. Not very sophisticated. The intent of Clause 13.7 is clear. But it remains to be seen whether a court would classify a notification as inadequate if it were contained in an email with several other communications. Certificates are also mentioned in Clause 13.6, namely to whom they are issued.

Clause 13.8 gives the *Project Manager* quite some authority. The *Project Manager* may withhold acceptance of any submission from the *Contractor*. This has no qualification, no ifs or buts. The *Project Manager*, as the *Client's* representative, must have this authority and it should be used carefully. Where acceptance is withheld for a reason not recognised by the contract, it is likely that a compensation event will result.

> **Example 2.1**
>
> A *Contractor* submits a design package to the *Project Manager* for acceptance, on time and in full compliance with the Scope. The *Client* is considering a reconfiguration to part of the *works* which will affect the areas that are the subject of the *Contractor's* design submission.
>
> The *Project Manager* does not accept the design, even though it appears fully compliant with the contract's current requirements. A few days later the *Project Manager* can instruct a change to the Scope and the *Contractor* must amend the design previously prepared and submitted.
>
> The *Project Manager's* actions will create a compensation event (Clause 60.1(9)), but the impact will be much less than had the original design submission been accepted and the *Contractor* allowed to place orders for Materials, subcontracts etc. and potentially start construction work.
>
> Here, the authority given to the *Project Manager* has enabled the design changes to be managed effectively and the cost and time impacts to be minimised.

This clause applies to all the submissions made to the *Project Manager*, for example programmes (Clauses 31.1 and 32.2), compensation event notifications (Clause 61.3), quality documents (Clause 40.2), a proposal for an alternative guarantor (X4.2) etc.

Submissions made to the *Project Manager* and *Supervisor* usually require a response. That response is often to accept, or to not accept. The act of accepting a communication is covered in Clause 14.1. By accepting a submission, the *Project Manager* or *Supervisor* is not affecting the liabilities of the *Client* and *Contractor*. There is one exception to this; the acceptance by the *Project Manager* of a programme does place responsibility on the *Client* for its own conduct and that of Others.

Communications accepting or not accepting submissions should be clear and should follow the basic rules set out in Clause 13 generally. Parties may disagree on the decision made by the *Project Manager* or *Supervisor*, but they have a better opportunity to resolve their disagreement if the reasons for it are clear. Too often we see disputes concerning the silence of one of the participants which, in time, leads to frustration.

As with most standard forms of contract, the ECC allows the delegation of authority. The *Project Manager* and *Supervisor* can delegate any of their actions; there is no limit provided in the ECC. But people or firms undertaking these roles need to be conscious of any limitations in their contracts with the *Client*. Those persons identified as *Project Manager* and *Supervisor* remain responsible for the actions of their delegates. Once again, clear communication is needed with this subject. The ECC requires communications about delegation to be notified, so Clause 13.7 applies. The *Contractor* is entitled to know who will be issuing instructions and to whom communications must be issued. Delegation can occur for several reasons; the size or geographical diversity of the *works*, the need for specialist resources or just the short-term absence of people, owing to holidays or training courses. The failure to identify who can do what on a construction project can and does lead to disputes, particularly where a delegate's actions, with hindsight, prove to be unhelpful.

The replacement of the *Project Manager* or *Supervisor* is covered by Clause 14.4. The *Client* may do this provided that the *Contractor* is notified. This clause is only used where the named person in the Contract Data is changed. For communications to be effective, the *Contractor* must know the name and contact details of the recipient(s).

17. Clause 14.3

Clause 14.3 gives the *Project Manager* the authority to instruct changes to the Scope or a Key Date. Such instructions typically, but not always, lead to compensation events under Clauses 60.1(1) and 60.1(4), respectively, which may change the Prices, and Key Dates or the Completion Date. Note that the *Project Manager* does not otherwise have the authority to change any other aspect of the contract, such as the Contract Data, Bills of Quantities, Activity Schedule etc.

Disputes about such instructions are common for several reasons.

- The *Project Manager* claims that the instruction was merely requiring the *Contractor* to do something that was already an obligation, so no compensation event is due.
- The *Project Manager* denies having made the instruction, although this can be avoided by following the rules in Clause 13.1. Interestingly, Clause 14.3 does not refer to a notification so Clause 13.7 does not apply. We would always recommend that such instructions are issued separately.
- Where an instruction has been identified, its requirements are unclear. If the *Client* has ensured that the Scope is written clearly and unambiguously, this will soon be undone by a series of poorly worded instructions from the *Project Manager*.

So, we need the *Project Manager's* instructions changing the Scope or a Key Date to be written clearly. But this also applies to any response to queries from the *Contractor*, Subcontractors or suppliers. Changing a Key Date is presumably straightforward; the date changes from X to Y. Clause 14.3 doesn't appear to offer the *Project Manager* the authority to change the *condition* attached to a Key Date.[iv] But changing the Scope requires careful consideration and drafting. Ideally the instruction should identify which part of the existing Scope is being amended or replaced and the new description that will be inserted. The language and style should replicate the original text so that the new obligations dovetail with the pre-existing ones.

18. The importance of communication in information modelling and Clause X10

Secondary Option X10 was introduced into the ECC for the first time in 2017 with the publication of the NEC4 suite. Reflecting a change in the way business is undertaken in the industry is a key driver of revising standard forms. The increasing use of Building Information Modelling (BIM) required the contract to be adapted. Clause X10 is designed to be used alongside more detailed BIM requirements; it acts as a bridge between the ECC and those detailed requirements that will typically be contained in the Scope.

NEC has produced a practice note[v] on this subject which suggests:

> NEC4 secondary Option X10 has been designed to work on a protocol independent basis. That is, there is no requirement to include a particular protocol such as that published by CIC; the requirements for an Information Model can all be included in the *Client's* Information Model Requirements in the Scope and the *Contractor's* Information Execution Plan.

However, the Winfield Rock report[vi] commented:

> Given the brevity of the NEC4 Option, it may be worthwhile looking at a standard BIM Protocol, and considering what other terms could or should be inserted into the Z clauses and/or [the Scope].

There are several ways in which detailed BIM requirements may be incorporated into an ECC contract. The use of the Construction Industry Council BIM Protocol[vii] is one way of doing this. The practice note[viii] says:

> The Protocol can be incorporated into the ECC by the selection of secondary Option X10 and the inclusion of the Protocol as the Information Model Requirements in the Scope.

BIM requirements in construction contracts will continue to evolve as the technology and the nature of its use evolve. As with many other aspects of the NEC contracts, secondary Option X10 requires close collaboration and communication between the various participants.

For example, see Clause X10.2:

> The *Contractor* collaborates with other Information Providers as stated in the Information Model Requirements.

There is an additional obligation[ix] to give an early warning in respect of matters that could adversely affect the Information Model.

The *Contractor* must submit an Information Execution Plan to the *Project Manager* at the start of the project and update it repeatedly throughout. The *Project Manager* must accept, or not accept, each submission. These requirements place additional communication obligations on the *Contractor* and the *Project Manager* and must be undertaken correctly. The nature of human beings' interactions with electronic data tends to increase the chance that an error will not only be missed but may be expanded on if not noticed.

The *Contractor* must obtain rights from Subcontractors[x] and this must be properly documented when forming the subcontract, presumably something that the *Project Manager* will want to check when exercising authority[xi] in the appointment of Subcontractors.

Clause X10 ends with some provisions on liability and clearly anticipates that this will be an important issue. The *Contractor's* liability is limited to the skill and care normally used by professionals undertaking such activities; in a dispute scenario much will turn on whether the *Contractor* and the *Client* can demonstrate whether this standard has, or has not, been achieved. Good quality communications between the *Contractor* and the *Project Manager* will support the avoidance of disputes and will prove important if needed as evidence in dispute resolutions.

19. The importance of communications in disputes

Communications are at the heart of disputes.

- Poor communications frequently cause, or contribute to, disputes.
- Past communications form part of the evidence presented in proceedings.

- Evidence of the crystallisation of a dispute typically involves communication or, occasionally, the omission of communication.
- Parties need to communicate when starting dispute resolution.
- Communications with each other and with neutral third parties form a considerable part of the evidential burden.
- Follow-up documents to dispute resolution (for example, a notice of dissatisfaction) are crucial documents, which, if issued incorrectly, could affect a Party's entitlement.

For all these reasons, communications must be considered both strategically and tactically by the Parties and not merely as an administrative action.

The crystallisation of a dispute is always a condition precedent to dispute resolution and the ECC is no different in this regard.

> Clause W2.2(1)
>
> A dispute arising under or in connection with the contract is referred to and decided by the *Adjudicator*.

With this clause, the *Adjudicator* has no jurisdiction to decide a dispute unless one exists. We must differentiate between disputes and the normal disagreements and differences of opinion that exist in commercial life. The English case of *Amec* v *Secretary of State for Transport*[xii] established the precedent in England and Wales and although it won't be binding in other jurisdictions, it sets out some commonsense explanation of what is, and possibly what is not, a dispute.

1. The word 'dispute' which occurs in many arbitration clauses and also in section 108 of the Housing Grants, Construction and Regeneration Act 1996, as amended,[xiii] should be given its normal meaning. It does not have some special or unusual meaning conferred upon it by lawyers.
2. Despite the simple meaning of the word 'dispute', there has been much litigation over the years as to whether or not disputes existed in particular situations. This litigation has not generated any hard-edged legal rules as to what is or is not a dispute. However, the accumulating judicial decisions have produced helpful guidance.
3. The mere fact that one party (whom I shall call 'the claimant') notifies the other party (whom I shall call 'the respondent') of a claim does not automatically and immediately give rise to a dispute. It is clear, both as a matter of language and from judicial decisions, that a dispute does not arise unless and until it emerges that the claim is not admitted.
4. The circumstances from which it may emerge that a claim is not admitted are Protean. For example, there may be an express rejection of the claim. There may be discussions between the parties from which objectively it is to be inferred that the claim is not admitted. The respondent may prevaricate, thus giving rise to the inference that he does not admit the claim. The respondent may simply remain silent for a period of time, thus giving rise to the same inference.
5. The period of time for which a respondent may remain silent before a dispute is to be inferred depends heavily upon the facts of the case and the contractual structure. Where the gist of the

claim is well-known and it is obviously controversial, a very short period of silence may suffice to give rise to this inference. Where the claim is notified to some agent of the respondent who has a legal duty to consider the claim independently and then give a considered response, a longer period of time may be required before it can be inferred that mere silence gives rise to a dispute.

6. If the claimant imposes upon the respondent a deadline for responding to the claim, that deadline does not have the automatic effect of curtailing what would otherwise be a reasonable time for responding. On the other hand, a stated deadline and the reasons for its imposition may be relevant factors when the court comes to consider what is a reasonable time for responding.

7. If the claim as presented by the claimant is so nebulous and ill-defined that the respondent cannot sensibly respond to it, neither silence by the respondent nor even an express non-admission is likely to give rise to a dispute for the purposes of arbitration or adjudication.

Amec Civil Engineering Ltd v *The Secretary of State for Transport*

The seven propositions put forward by HHJ Jackson (as he then was) in this case all have some connection to the Parties' communications. It is common for adjudicators to be challenged by the responding Party that they have no jurisdiction; one of the issues relied on is often that there is no crystallised dispute. It is usually correspondence, as evidence of the Parties' previous dealings, that can demonstrate the existence of a crystallised dispute. In ECC disputes, the existence of a dispute, more often than not, is shown by the *Contractor* and *Project Manager* disagreeing on the assessment of compensation events. Disputes over quality will draw in the *Supervisor* too.

Project Managers and *Supervisors* should note that *Clients* will be reluctant to defend proceedings brought by *Contractors* if the necessary actions of the *Client's* representatives cannot be evidenced.

20. Conclusion

This chapter has described the communication requirements in the ECC. We have emphasised how good communications should reduce the possibility of disputes arising. Where they do arise, a Party's past communications will have a significant bearing on its chances of success in dispute resolution proceedings.

NOTES

i The exception is Clause 12.3.
ii *Trant Engineering Limited* v *Mott MacDonald Ltd* [2017] EWHC 2061 (TCC).
iii Clause 10.1.
iv This could be achieved, though, by agreement of the Parties under Clause 12.3.
v Higgins P, Heaphy I and Croft A (2017). *NEC4 Practice Note 2. How to Use the CIC BIM Protocol with NEC4*. NEC, London, UK.
vi Winfield M and Rock S (2018) *Overcoming the Legal and Contractual Barriers of BIM*. UK BIM Alliance, London, UK.
vii Beale & Company (2018) *Building Information Modelling Protocol*, 2nd edn. Construction Industry Council, London, UK.
viii See note v, above.
ix Clause X10.3.
x Clause X10.6.

xi Clause 26.3.
xii *Amec Civil Engineering Ltd* v *Secretary of State for Transport* [2005] EWCA Civ 291.
xiii HMG (Her Majesty's Government) (1996) Housing Grants, Construction and Regeneration Act 1996. The Stationery Office, London, UK. Amended by HMG (2009) Local Democracy, Economic Development and Construction Act 2009. The Stationery Office, London, UK.

Chapter 3
Early warnings

21. What are early warnings and why do we need them?

Early warnings are almost a two-faced part of all NEC contracts. Designed primarily as a collaborative tool, they feature heavily in many disputes. The interests of balance mean that we should highlight their role in preventing disputes too. While early warnings assist the Parties' collaborative efforts, they also present clear sanctions to the *Contractor* in the event of non-compliance with the contract.

Let's look at what the contract requires in Clauses 15.1 and X12.

> **Clause 15.1**
>
> The *Contractor* and the *Project Manager* give an early warning by notifying the other as soon as either becomes aware of any matter which could
>
> - increase the total of the Prices,
> - delay Completion,
> - delay meeting a Key Date or
> - impair the performance of the *works* in use.
>
> The *Project Manager* or the *Contractor* may give an early warning by notifying the other of any other matter which could increase the *Contractor's* total cost. The *Project Manager* enters early warning matters in the Early Warning Register. Early warning of a matter for which a compensation event has previously been notified is not required.

Where secondary Option X12, on multiparty collaboration, is used, a further obligation to give early warnings is included. This is for the benefit of the wider group of Partners. The term 'Partner' includes the *Client* and the *Contractor*. In this respect, the *Contractor* needs to give any early warning to the *Client* or to another organisation, but not to the *Project Manager*.

> **Clause X12.3(3)**
>
> Each Partner gives an early warning to the other Partners when it becomes aware of any matter that could affect the achievement of another Partner's objectives stated in the Schedule of Partners.

Early warnings given under Clause X12 differ from those given under Clause 15.1. There is no obligation, unless the Partners agree to it, for an Early Warning Register or an early warning meeting in Clause X12. The remainder of this chapter refers to the procedures in Clause 15.

The contract is precise over the type of event that should provoke an early warning, when an early warning should be issued and by whom. The aim of the process is to highlight matters that present a risk to the project or to the Parties. The three key constituents of success in any project are cost, time and quality. The four items listed in Clause 15.1 reflect these, albeit using defined terms. The clause only requires an early warning of something that <u>increases</u> the Prices, not just merely changes them. The change to the Prices when they reduce does not trigger the contractual requirement to give an early warning, even though that presents a risk to both Parties.

In some respects, the contractual obligation to collaborate may seem odd. But because the *Contractor* and the *Project Manager* are required to behave in a specified manner, they know what to expect of one another. By becoming aware of a matter that might otherwise not have been apparent, the recipient of an early warning may then take action to mitigate or avoid the problem. If the recipient chooses not to take action, one of the Parties, if not both, is likely to suffer the consequences of not acting on such a decision, which are likely to be additional cost and delay, both breeding grounds for disputes.

Clause 15.1 also contains a rare example of an optional task in an NEC contract. The *Project Manager* and the *Contractor* <u>may</u> give an early warning of something that could increase the *Contractor's* costs. When working with main Options C, D, E and F, an increase in the *Contractor's* costs will feed through into increasing the *Client's* costs too. But the obligation of the *Contractor* to provide cost forecasts to the *Project Manager* in these Options means that the *Project Manager* will be aware of these matters; therefore, a separate early warning isn't usually necessary. The ability to give an early warning in these circumstances is helpful to both Parties, but particularly the *Contractor*.

We look at the sanctions for non-compliance later in this chapter in more detail. Briefly, for the *Contractor*, they are the potential reduced assessed value' of a compensation event and the disallowing of Defined Cost, both in the situation where an early warning was not given but could have been. There is no specified sanction for the *Client*, but a failure of the *Project Manager* to issue an early warning is likely to lead to additional cost or delay for the *Client*.

The sanction for the *Contractor* is set out in the contract, but there is no sanction set out for the *Client*. Therein lies the incentive for *Contractors* to be 'trigger-happy' in their notification of early warnings. 'Best send it, just in case…' is a refrain heard in *Contractors*' offices around the world. Some early warnings are not well received in these circumstances and this can have an unhelpful impact on relationships. Early warnings, or the lack of them, are rarely the sole cause of a dispute. More usually they are a component of a wider problem in the project.

Early warnings do not have to be accepted. Or agreed. Or approved. Or any other such word. If the recipient does not wish to react to the communication then, subject to any early warning meeting or administration by the *Project Manager* of the Early Warning Register, it may be ignored. A common feature of adjudications on NEC contracts is the sight of documents where someone has said 'I don't accept this' or 'your early warning is rejected'. Such statements are pointless and only indicate an ignorance of the contract's requirements. If you receive what you consider to be an unmerited early warning, it is best to look at what you're being told and, if the matter is pertinent, deal with it.

The final sentence of Clause 15.1 is helpful in avoiding duplication of communications. Not all compensation events are preceded by an early warning and not all matters that require an early warning turn into compensation events. But where a compensation event occurs immediately, with little or no warning, it is unnecessary for the *Project Manager* or the *Contractor* to notify an early warning. Where it is known that a compensation event has occurred, the relevant person should issue a notification. An early warning in these circumstances will add little to the Parties' knowledge.

22. What should happen once an early warning has been given?

Early warnings that are properly given should lead to the *Project Manager* and the *Contractor* considering the matters warned about. Not all early warnings will require the formalities of an early warning meeting, as we describe later. But all will require some level of consideration, even if that just leads to a quiet dismissal of the subject. All early warnings must be entered in the Early Warning Register by the *Project Manager*. This provides a record of the discussions and what the *Project Manager* and the *Contractor* have agreed to do.

Before looking at the details of the register and the early warning meetings, let's see what the contract requires.

> **Clause 15.2**
>
> The *Project Manager* prepares a first Early Warning Register and issues it to the *Contractor* within one week of the *starting date*.

The first version of the Early Warning Register must be prepared by the *Project Manager* within one week of the *starting date*. Given the importance of effective contract administration in avoiding disputes, this is an important task, even if there is little to include at this stage. The two entries from the Contract Data (parts one and two) must be included to reflect any issues that the Parties identified at tender stage. By having the register set up and ready to use, the *Project Manager* is starting the process of risk management for the project and enabling it to add value to the Parties. We occasionally come across projects where the *Project Manager* has delegated these duties to the *Contractor*, as if to dilute the importance of these tasks to administrative minutiae. *Project Managers* who do this are in breach of Clauses 10.1 and 15.2, in addition to their own contractual obligations to the *Client*. The early warnings processes are important and the various obligations in the contract should not be toyed with.

Not all early warnings require an early warning meeting, either immediately or at all. Many projects have a routine (say monthly) early warning meeting or include early warnings on the agenda for other, more general, meetings. But the contract requires a regular pattern of meetings and for these to start promptly after the *starting date*.

> **Clause 15.2 (continued)**
>
> The *Project Manager* instructs the *Contractor* to attend a first early warning meeting within two weeks of the *starting date*.

The clause goes on to say that later meetings are held ad hoc at the instruction of either the *Contractor* or the *Project Manager* or at the interval stated in Contract Data part one. So here we have the contract once again prescribing how the Parties should collaborate. By demanding the first register and the first meeting within defined timescales (as with most processes in NEC contracts), the contract is ensuring that the structure of risk management is put in place at the start and therefore, we hope, makes it more likely that the Parties will use it and therefore less likely that any matters will lead to dispute.

Early warning meetings can be attended by other people and by Subcontractors if this will be helpful; our experience is that the meetings will generally succeed if there is a wider attendance. If, for example, the matter being discussed involves the work of a specialist Subcontractor, the attendance of that Subcontractor and possibly the *Client's* designers will produce a more effective outcome than if the only attendance was by the *Project Manager* and the *Contractor*, acting in proxy for Others.

Clause 15.3 explains what happens at the early warning meeting. It doesn't require merely the attendance of people but their 'co-operation'. There is a vital difference between the two actions; demanding the co-operation of attendees is a positive requirement that, if delivered, will provide a greater opportunity for avoiding or mitigating the problem under discussion. The product of the co-operation should be

- proposals for avoiding or reducing the risks posed by the matter(s) in question
- seeking solutions
- deciding on actions
- deciding whether to remove 'closed' matters from the register
- reviewing previous actions.

Clause 15.4 then places further requirements on the *Project Manager*.

Clause 15.4

The *Project Manager* revises the Early Warning Register to record the decisions made at each early warning meeting and issues the revised Early Warning Register to the *Contractor* within one week of the early warning meeting. If a decision needs a change to the Scope, the *Project Manager* instructs the change at the same time as the revised Early Warning Register is issued.

So again, we have timed obligations for the *Project Manager*, this time at the conclusion of the meeting. The issuing of any instruction in combination with the revised register should minimise the chance of one of these being erroneously omitted.

23. The early warning process in practice

The early warning process is prescribed in the contract and we have explained earlier how it is supposed to work. But, as with many contractual processes, it requires some skill and judgement in its application. We have seen these symptoms in projects, which can suggest a lack of judgement.

- The number of early warnings notified by the *Contractor* dwarfs the number issued by the *Project Manager*.
    - Could this mean that the *Contractor* is 'trigger-happy'?
    - Could this mean that the *Project Manager* lacks understanding of the contract?
- Early warnings and compensation event notifications are used interchangeably by both Parties, but the lack of a compensation event notification is then seized on by the *Project Manager* or *Client* in an attempt to avoid payment.
- A refusal to attend early warning meetings, either overtly or by prevarication.
- Giving early warnings for trivial matters that shouldn't occupy the decision-making processes in either Party – for example, a vague newspaper report that the long-term weather report is for a wet spring or a dry summer etc. Where a report like this isn't backed up by a professional forecast, what can be done to react to it?
- On one occasion, we saw an instruction from the *Project Manager* to the *Contractor* requiring an early warning to be given by the *Contractor* to the same *Project Manager*!
- While early warnings should be given as soon as becoming aware of the matter, issuing a written warning can usually wait until later in the day. If something is truly urgent, a telephone call should be made.

Understanding when to issue an early warning is partly a process of understanding the contract and also one of understanding the other people you are working with (or co-operating with?) on your project. It is surely common sense that matters of common interest that could affect the success of the project should be discussed openly. The provision of contractual obligations in this respect merely formalises what the Parties should be doing anyway.

Disputes in construction projects tend to involve cost and time overruns. The early warning processes, if operated properly, will minimise the chances of such overruns and therefore minimise the chances of disputes occurring.

24. What happens when the *Project Manager* and the *Contractor* don't comply with the early warning processes?

The simple answer to this question in respect of the *Project Manager's* failure is that the project is likely to take longer or cost more money for the *Client*, or both. Where the *Project Manager* is aware of a suitable matter that will have an impact on cost, time or quality and chooses not to share this with the *Contractor*, this will, in many situations, lead to a higher bill for the *Client*. But there are not express sanctions in the contract explaining what will happen if the *Project Manager* fails to act as stated in Clause 15.

Sanctions for the *Contractor's* failure to give an early warning when one should have been given are clearly set out. They sit in two places.

- In all main Options, the assessment of compensation events will be undertaken differently, to accommodate the *Contractor's* failure to give an early warning.
- In main Options C, D, E and F, any costs incurred only because the *Contractor* did not give an early warning will be disallowed (i.e. will not be reimbursed by the *Client*).

These provisions are important and frequently contribute to disputes so it's worth looking at them in detail, starting with Clauses 61.5 and 63.7.

> **Clause 61.5**
>
> If the *Project Manager* decides that the *Contractor* did not give an early warning of the event which an experienced contractor could have given, the *Project Manager* states this in the instruction to the *Contractor* to submit quotations.
>
> **Clause 63.7**
>
> If the *Project Manager* has stated in the instruction to submit quotations that the *Contractor* did not give an early warning of the event which an experienced contractor could have given, the compensation event is assessed as if the *Contractor* had given the early warning.

The language in these clauses reflects the subjective aspects of giving early warnings. The sanction is for the failure to give an early warning when one <u>could</u> have been given. Not 'should', but 'could'. The test is whether an experienced contractor, as distinct from the *Contractor*, could have given one. We suggest that the chances of getting caught by this clause are quite high and that this therefore encourages the trigger-happy nature of some *Contractors*. The *Project Manager's* obligation is not optional; it is something that must be done when reading Clauses 10.1 and 61.5 together.

The implication of Clause 61.5 is that the *Contractor* must allow for this in its quotation and, should the quotation not be accepted, the *Project Manager* must do likewise when making any assessment. The clause has a history of confusing first-time users and of creating arguments between the Parties. After all, even if an early warning could have been given, there might have been no impact on cost, time or quality. We demonstrate the issues in two examples from real life adjudications.

> **Example 3.1**
>
> A *Contractor* is working to refurbish a railway station under the ECC with main Option A. Part of the work involves installing prefabricated steel bicycle racks in the station car park prior to resurfacing the entire car park.
>
> While excavating to install the concrete foundations for the new bicycle racks, the *Contractor* uncovers a previously buried concrete floor slab, which must have accommodated an old station building many years previously. The *Contractor* spent £47,000 excavating and removing the concrete slab before installing the new (concrete) foundations for the bicycle racks.
>
> The bicycle racks were duly installed, the car park resurfaced and the finished product looked excellent. Seven weeks after discovering the concrete, the *Contractor* notified the *Project Manager* of a compensation event – that difficult physical conditions had been encountered and sought the recovery of the £47,000 that the excavation had cost.
>
> The *Project Manager* made a notification under Clause 61.5 saying that an early warning could have been given by the *Contractor* (the £47,000 increase in the Total of the Prices) and that it wasn't given.

Therefore, the compensation event should be assessed as if an early warning had been given. The *Project Manager* explained that had it been made aware of the problems, the design would have been changed to situate the bicycle racks elsewhere, where the physical conditions were more favourable, and that the £47,000 additional excavation costs would not then have been incurred. Therefore, the *Contractor* was not due any additional payment.

In Example 3.1, we see how the early warning provisions have teeth and the *Contractor* was unable to recover the money that was needlessly spent. Our second example is one where the failure to give an early warning was immaterial.

Example 3.2

A *Contractor* was working on the construction of a high-voltage electrical substation under the ECC with main Option B. Other contractors, working for the same *Client*, were also present on the Site. The *Contractor* in this example was instructed by the *Project Manager* to provide people and Equipment to assist one of the other contractors (who we shall call 'Contractor 2' for the purposes of this question).

The *Contractor* notified a compensation event to the *Project Manager*, correctly pointing out that the instruction had changed the Scope. The *Project Manager* sought to use the provisions of Clauses 61.5 and 63.7 to reduce the amount due as a result of this work. But the *Contractor* pointed out that its assistance to Contractor 2 was known to the *Project Manager* as it resulted from one of the *Project Manager's* own instructions. An early warning would not have told the *Project Manager* anything new and consequently the compensation event assessment should not be amended as the assessment would have been the same with an early warning as it was without.

Our two examples have referred to the situation where the failure to notify an early warning affects the assessment of compensation events. In all main Options, this failure affects the Total of the Prices. In main Options C, D, E and F, the failure also has an impact on the Price for Work Done to Date through the process of Disallowed Cost. This is defined in Clause 11.2(26) (for main Options C, D & E) and 11.2(27) (for main Option F).

Clauses 11.2(26) and 11.2(27)

Disallowed Cost is cost which […]

- was incurred because the *Contractor* did not […]
  - give an early warning which the contract required it to give.

Notice the slightly changed language here. In Clauses 61.5 and 63.7, the failure of the *Contractor* is one where an early warning <u>could</u> have been given. Here the failure is to give an early warning when the contract required it. This, we would suggest, omits early warnings that the *Contractor* <u>may</u> make in respect of its costs from these sanctions.

Disallowed Cost is deducted from Defined Cost in the *Project Manager's* assessment of the Price for Work Done to Date; therefore, the failure to give an early warning will have an impact on the regular payments received by the *Contractor*. We describe the payment processes for main Options C, D, E and F later but it's important to understand here that the *Client* cannot be expected to reimburse costs that were unnecessarily incurred by the *Contractor*. This is what Clause 11.2(26) sets out to avoid.

25.   How does the failure to give early warnings affect the Parties' relationship?

We have explained how the failure of either the *Contractor* or the *Project Manager* to give early warnings can have an impact on cost, time or quality. It is a fact of commercial life in any industry that differing Parties will view the reasons for these problems in different ways. Where one Party holds the other one responsible for a situation that could, might or would have been prevented with an early warning, or not, as the case may be, a dispute will not be far away. Once again, our advice is to do what the contract says as the primary way of avoiding trouble.

Chapter 4
Programme

26. When is the first programme required?

The programme is central to almost all processes in the ECC. Understanding its role and doing what the contract requires increases the Parties' chances of a successful project many times over. The contract allocates the duty of preparing and updating the programme to the *Contractor*, but the intention is for a tool that both Parties may rely on in planning the various activities in the project.

The contract provides for a programme, it would appear however good or bad, to be included in Contract Data part two by the *Contractor*. There are two optional statements concerning time and programme in the Contract Data part two

- identification of a programme
- identification of the *completion date* for the whole of the works.

These provisions are aimed at including time and programme issues in a competitive tendering scenario, although this is not clearly stated. Many clients want to see a programme with tenders; many more do not. Clause 31.1 sets out what happens if a programme has not been identified in Contract Data part two.

> **Clause 31.1**
>
> If a programme is not identified in the Contract Data, the *Contractor* submits a first programme to the *Project Manager* for acceptance within the period stated in the Contract Data.

The period stated in the Contract Data can be decided by the Parties (but more probably by the *Client*) but it needs to be short; the *Contractor* should have prepared a programme at tender stage even if it was not submitted with Contract Data part two. The early submission of the programme to the *Project Manager* will enable the necessary dialogue to begin and enable each to understand the other's requirements. This type of dialogue is essential in the avoidance of disputes. A good programme submission will provoke discussion and identify potential clashes and problems in advance of them occurring. The contents of the programme required in Clause 31.2 ensure that such issues will be highlighted early on.

The contract contains a disincentive to poor management in Clause 50.5; the *Project Manager* must certify the retention of 25% of the Price for Work Done to Date until the first programme is submitted for acceptance by the *Contractor*, if one isn't identified in Contract Data part two. This acts to encourage the *Contractor* to make a timely submission. We regularly see a Z clause change to the contract making this clause effective whenever a programme is due (typically monthly) for submission by the *Contractor*.

Generally, this isn't helpful. The purpose of Clause 50.5 as it is currently written is to ensure that the *Contractor* can produce a compliant programme. Once this has been achieved, it may be referred to as a benchmark in future submissions if their quality falls short. Where we see Z clause amendments in this regard, they tend to feature quite highly in problem projects, a minor symptom of an approach to contract management that is less than collaborative. The requirement to retain the 25% falls away when a programme is submitted 'showing the information which the contract requires'. In other words, the programme must comply with Clause 31.2. There is no requirement for the programme to be accepted by the *Project Manager* to lift the retaining of the 25%. The 25% often causes people to raise their eyebrows but this figure is somewhat less than the 50% found in earlier versions of ECC. The best advice to the *Contractor*, as usual, is to follow the contract precisely, in particular Clause 31.2.

27. The terminology of time and programmes in the ECC

The *Contractor's* primary obligation, in respect to time, is given in Clause 30.1.

> **Clause 30.1**
>
> The *Contractor* [...] does the work so that Completion is on or before the Completion Date.

The ECC contains several terms for use in time and programmes; it is helpful to set some of them out here before we explain how they are used.

Contract Date	Defined in Clause 11.2(4) as 'the date when the contract came into existence'. Given the Construction industry's propensity for forming contracts without signatures or any other identifiable means of providing agreement, this broad definition may become a part of disputes. We occasionally see disputes where the date of forming the contract becomes important.
starting date	This isn't a defined term, but it is identified in Contract Data part one. By not being defined, the phrase has its ordinary meaning. The contract also uses the phrase '*access date*' so the use of '*starting date*' clearly reflects the possibility that the works may start offsite first, for example with design or offsite fabrication.
access date(s)	Also identified in Contract Data part one. The use of more than one *access date* would be appropriate for multisite projects or those with a phased provision by the *Client*.
Key Dates	Key Dates are defined in Clause 11.2(11) and their use is explained in Clause 25.3. We describe their use in more detail later but, in brief, they provide a mechanism for the *Client* to co-ordinate the activities of the *Contractor* with other suppliers and for the *Client* to avoid paying the costs of the *Contractor's* time-related failures. Another example of an NEC term that is both defined and identified, the initial Key Dates are listed in the Contract Data part one as *key dates* but they change with compensation events, acceleration, accepted Defects and instructions notified by the *Project Manager*.

Completion	Defined in Clause 11.2(2) as the required state of the *works* as set out in the Scope or otherwise determined by the *Project Manager*.
Completion Date	Defined in Clause 11.2(3) as the *completion date* identified in Contract Data part one. Another example of an NEC term that is both defined and identified, the initial Completion Date is listed in the Contract Data part one as the *completion date* but it changes with compensation events, acceleration and accepted Defects. Where secondary Option X5 is used, the *completion date* for each section must also be identified in Contract Data part one.
planned Completion	A phrase that is neither defined nor identified so it has its ordinary meaning. Probably the most valuable piece of information for the *Client* during any project is the expected date when the *works* will be complete. The *Contractor* must forecast this date and specify it in all programme submissions.
take over	Take over isn't defined but is included as a way of differentiating between the works being complete and the *Client* starting to use them. Clause 35 places certain requirements on the *Contractor*, the *Client* and the *Project Manager*.

28. What is an Accepted Programme?

The term 'Accepted Programme' is used throughout the ECC. The contract requires the *Contractor* to keep the programme up to date with reality and not just merely refer back to a 'day one' or 'baseline' programme, which, with every passing day, will inevitably become less helpful. The Accepted Programme is defined in Clause 11.2(1):

> The Accepted Programme is the programme identified in the Contract Data or is the latest programme accepted by the *Project Manager*. The latest programme accepted by the *Project Manager* supersedes previous Accepted Programmes.

The role of the *Project Manager* is obviously important here. The *Project Manager* assesses programmes submitted by the *Contractor* and must do this within defined timescales, which we describe later. The reliance on the Accepted Programme will only work if both the *Contractor* and the *Project Manager* work together and within the timescales required.

29. What should programmes show?

Clause 31.2 sets out the requirements for a compliant programme submission. Several of these requirements go well beyond what might be accommodated on a Gantt chart and therefore point towards a wider documentary submission. For example

- provisions for health and safety requirements
- provisions for the procedures set out in the contract
- a statement of how the *Contractor* plans to do the work.

The final list item in Clause 31.2 is a 'cover-all', which requires the *Contractor* to provide whatever additional information is required for the Scope. Typically, we expect to see a specification by the software for preparing the programme, if the *Client* has a preference.

With the introduction of extended quality requirements in the NEC4 publications there is now a potential overlap between the need for statements of how the *Contractor* plans to do the work (Clause 31.2) and a quality plan (Clause 40.2). The *Contractor* should ensure consistency between these two submissions.

In common with many other areas of the contract, the programme provisions provide a good framework to encourage collaboration and the exchange of important information. Without the provision of this information, the efficient management of the project will suffer and, inevitably, so will relationships, bringing the possibility of disputes closer.

The requirements for dates, a float, time risk allowances and the like are made to ensure that the *Project Manager* (and therefore the *Supervisor*, *Client* and Others too) are aware of all the progress of the works and what is required and when.

The primary requirements of Clause 31.2 are to show when things will happen and how those things interact with other things. The contract isn't prescriptive about how this is depicted, although the *Client* is free to add specific requirements to the Scope. If the *Contractor* can show the order and timing of the various operations, the *Client*, *Project Manager* and *Supervisor* are more likely to know how they can support the progress of the project. If and when disputes occur, their resolution will be hampered by a paucity of information in the programme. Conversely, where the programme sets out the *Contractor's* plans clearly, less effort will be needed in resolving the dispute.

The float plays a big part in disputes. The usual question is, 'Who owns the float?' The ECC allows the *Contractor* to retain the benefit of a terminal float; it does this in Clause 63.5 when dealing with the assessment of compensation events.

> **Clause 63.5**
>
> A delay to the Completion Date is assessed as the length of time that, due to the compensation event, planned Completion is later than planned Completion as shown on the Accepted Programme current at the dividing date.

A similar provision is then made for the effect on Key Dates.

The effect of Clause 63.5 is to entitle the *Contractor* to additional time in respect of each compensation event where there is a time impact on planned Completion. The entitlement remains even if the *Contractor* is benefiting (or indeed suffering) from a terminal float (or a negative terminal float).

Time risk allowances are different from float and must be identified in the programme submission. The contract doesn't define this term, so it has its ordinary meaning.

Float, time risk allowances and planned Completion will play major parts in the successful management of projects and, conversely, in the resolution of disputes, once they occur. Therefore, the regular submission, by the *Contractor*, of revised programmes and the regular acceptance of them, by the *Project Manager*, are crucial.

30. What is the role of programmes in disputes?

Time issues form a significant part of almost all construction disputes. Time is money, as they say, and the delay of any single activity has the potential to cause knock-on delays to many other activities. The deployment of people and Equipment in construction projects usually costs money by the hour, day or week and therefore delay turns into cost. Those additional costs must be met by someone; those providing people and Equipment require paying regardless of the problems of the *Contractor* and the *Client*. So, participants to an ECC contract (and, indeed, any other construction contract) often find themselves debating who should pay these costs when something has been delayed or disrupted on a project.

Disputes in ECC contracts frequently revolve around compensation events, which we describe in more detail in Chapter 6. It can be argued that all the compensation events listed in Clause 60.1 (and those listed in the X and Y clauses) involve some element of time in that they delay or disrupt the work of the *Contractor*. The following events specifically refer to a time issue.

Clause 60.1(2)	The *Client* does not provide access to the Site by the later of the *access date* and the date for access shown on the Accepted Programme.
Clause 60.1(3)	The *Client* does not provide something that it is to provide by the date shown in the Accepted Programme.
Clause 60.1(4)	The *Project Manager* gives an instruction to stop or not to start any work or to change a Key Date.
Clause 60.1(5)	The *Client* or Others do not work within the times shown on the Accepted Programme.
Clause 60.1(6)	The *Project Manager* or *Supervisor* does not reply to a communication from the *Contractor* within the period required by the contract.
Clause 60.1(11)	A test or inspection conducted by the *Supervisor* causes unnecessary delay.
Clause 60.1(15)	The *Project Manager* certifies take over of a part of the *works* before both Completion and the Completion Date.
Clause 60.1(19)	An event stops the *Contractor* from completing the whole of the *works* by the date for planned Completion shown on the Accepted Programme.

These events can all be mitigated or removed by the efficient use of the programme. When they occur, the cross references to the Accepted Programme and to information such as the *access date* demonstrate the need for the contract to be prepared properly in the first place and then administered correctly. Where these things aren't done, the assessment of compensation events will prove more difficult than it needs to be, and disputes will arise as to the correct interpretation of the contract.

31. Keeping the programme up to date

The programme must be kept up to date; that much is clear just from a logical viewpoint. It is no use to either Party if their respective planning activities cannot rely on the main document designed for that purpose. The ECC requires the programme to be kept up to date. The revisions must be submitted

- when the *Project Manager* instructs the *Contractor* to do so
- at least as frequently as stated in the Contract Data part one
- with compensation event quotations.

A revision may be submitted

- when the *Contractor* chooses
- during discussions for, and implementation of, acceleration
- during discussions for, and implementation of, whole life cost proposals.

So, the revisions are needed (i) as a matter of routine and (ii) when 'something' is happening that may change the future sequence of events. In both situations, the contract requires revisions to include

1. the actual progress on each operation
2. how the *Contractor* intends to deal with delay
3. how the *Contractor* intends to deal with Defects
4. any other changes that the *Contractor* has made to the future sequence.

32. The authority of the *Project Manager* to make changes

The *Client's* side of any construction contract requires an element of control; to make certain things happen, to stop other things from happening at all. More commonly, in respect of time, it is to change the date or sequence when individual operations occur. The *Project Manager* is given the necessary authority to instruct the *Contractor* in Clause 34; to stop work, to not start work, to start work, to restart work or to remove work entirely from the Scope. This final action is consistent with the *Project Manager's* ability to change the Scope given in Clause 14.3.

Clause 34 doesn't require the *Project Manager* to notify the instructions, therefore the requirement in Clause 13.7 to issue the instruction(s) separately does not apply here. But we recommend that instructions of this gravity are issued separately and, of course, they must at least comply with Clauses 13.1 and 13.2. Assessing the impacts of several instructions to stop, start and reschedule works is a difficult enough activity at the best of times and will be made more so if there is not an effective audit trail of who instructed what, and when. Once disputes occur, and collaboration inevitably deteriorates, the chronological vacuum becomes filled with a combination of speculation and, in some situations, complete fantasy.

33. Access

The *Contractor* must have access to the Site on the dates promised in the contract. Without that clear (and, if specified, unfettered) access, progress is likely to suffer. Remember Clause 25.1.

> Clause 25.1
>
> The *Contractor* shares the Working Areas with Others as stated in the Scope.

The reverse of this requirement is that, where such requirements aren't stated in the Scope, the *Contractor* will assume that the Working Areas will be free of Others' activities and any change to this will create a compensation event.

While the meanings of 'Site' and 'Working Areas' differ slightly (see Clauses 11.2(17) and 11.2(20)), there is a similar statement in Clause 33.1.

> **Clause 33.1**
>
> The *Client* allows access to and use of each part of the Site to the *Contractor* which is necessary for the work included in the contract. Access and use is allowed on or before the later of its *access date* and the date for access shown on the Accepted Programme.

Clause 33 therefore emphasises the importance of getting information on to the programme and keeping it up to date. There appears to be little encouragement for the *Contractor* to 'offer' later access to the Site or to a part of it as that decision may not be reversed except with the *Project Manager*'s consent.

Keeping accurate, and hopefully agreed, records of access provision, and denial, is essential to the successful assessment of any compensation events that flow from access issues. Those records will also be useful in pursuing or defending any disputes that arise.

34. Take over

Take over does not always happen at Completion, hence the contract's separate provisions for take over. The *Client* is not required to take over the *works* before the Completion Date (Clause 35.1) even if they are complete, provided that this is stated in the Contract Data. Take over must occur within two weeks of Completion in other scenarios.

Take, for example, the construction of a school building with a Completion Date of 31 July, intended for occupation at the start of the academic year on 1 September. If the *Contractor* reaches Completion in, say, May, the school authorities will have no use for the building and will incur maintenance, security and insurance liabilities. So, it is in their interests not to take the building over early.

As another example, consider the development of a supermarket. If the buildings and external works are completed early, it is likely that the *Client* will wish to take over the *works* as soon as Completion is achieved to start earning profits on the development.

Both examples are straightforward. More issues arise when the *Client* takes over parts, or all, of the *works* before they are complete. Clause 35.2 allows the *Client* to do this.

In all situations, the *Project Manager* must certify the date on which take over occurs and the extent of that take over (Clause 35.3). Where the contract includes secondary Option X7 (delay damages), the situation can become further complicated because of an onerous duty placed on the *Project Manager*.

> Clause X7.3
>
> If the *Client* takes over a part of the *works* before Completion, the delay damages are reduced from the date on which the part is taken over. The *Project Manager* assesses the benefit to the *Client* of taking over the part of the *works* as a proportion of the benefit to the *Client* of taking over the whole of the *works* not previously taken over. The delay damages are reduced in this proportion.

This provision, while making sense in a logical way, is incredibly difficult to operate in real life. Take our earlier example of a supermarket under construction. Does your average *Project Manager* have the skills to assess the benefit to a supermarket owner of part of its store being taken over before Completion? How much does the petrol station earn in profits, compared with the bakery? Or should any partial take over be assessed on the basis of floor area? Add to the mix the age-old difficulties with the law on liquidated damages and you have a huge potential for dispute here. The only advice that we can offer is for both the *Contractor* and the *Project Manager* to keep accurate, agreed records for the time when such debates commence.

In this discussion, it is also worth considering the wording of Clause 60.1(15).

> Clause 60.1
>
> The following events are compensation events […]
>
> Clause 60.1(15)
>
> The *Project Manager* certifies take over of part of the *works* before both Completion and the Completion Date.

This means that there is a compensation event if the *Client* takes over early as that will inevitably disrupt the *Contractor* and the time originally agreed by the Parties to achieve Completion will not have been provided. Where the take over happens after the Completion Date, there is not a compensation event, as the *Client* in these circumstances is effectively mitigating any losses that occur from the *Contractor's* breach.

35. Acceleration

Acceleration, we hope, is one area of the contract that should generate few disputes, as there is no compunction on the *Contractor*. The *Contractor* is not obliged to accelerate the *works* to achieve Completion before the Completion Date, nor is there an obligation to prepare a quotation. If the *Contractor* and the *Project Manager* are prepared to consider such a change, Clause 36 sets out the process to be followed. To minimise the chance of disputes occurring, the various administrative requirements of Clauses 36.2 and 36.3 should be followed by both the *Contractor* and the *Project Manager*.

36. Key Dates

Key Dates appear in Clause 25.3, separately from the section on time.

> **Clause 25.3**
>
> If the *Project Manager* decides that the work does not meet the Condition stated for a Key Date by the date stated and, as a result, the *Client* incurs additional cost either
>
> - in carrying out work or
> - by paying an additional amount to Others in carrying out work
>
> on the same project, the additional cost which the *Client* has paid or will incur is paid by the *Contractor*. The *Project Manager* assesses the additional cost within four weeks of the date when the Condition for the Key Date is met. The *Client's* right to recover the additional cost is its only right in these circumstances.

As with any dispute avoidance advice elsewhere, we start with the need to ensure that the Key Dates and conditions are properly set out in the contract. The Key Dates start with *key dates* in the Contract Data part one, recognising that the *key dates* may change during the validity of the contract with instructions from the *Project Manager* and compensation events. The language should match the usual NEC style and describe in objective terms what the *Contractor* must achieve.

Clause 25.3 refers to a Condition (emphasis added) but this word is not defined in the contract. In Contract Data part one, the word used is *condition*, i.e. something that is not defined but is identified. There may be a reason why this apparent inconsistency exists, but a well-worded Z clause could remove it entirely. It seems to us that either a definition of Condition or the amendment in Clause 25.3 to *condition* is required.

However the words are used, the dates and the conditions must be set out clearly in the Contract Data part one and must then be shown on all programmes.

Clause 14.3 allows the *Project Manager* to instruct a change to a Key Date; this in turn creates a compensation event via Clause 60.1(4).

The sanctions available to the *Client* for the *Contractor's* failure to meet a *condition* on time are stated. They effectively exclude what we might call consequential loss. It is also an exclusive remedy, albeit an unascertained one. The purpose of the clause is to assist the *Client* in co-ordinating multiple suppliers on a single project and providing financial protection against some of the costs of doing so. The clause allows the *Client* to recover its own costs as well as any costs paid to Others. For example, the *Contractor* may be required to install cabling and other ancillary ICT infrastructure before another supplier, or the *Client* itself, can install the hardware that will be connected by the infrastructure.

The correct working of Clause 25.3 requires the *Project Manager* to assess the *Client's* additional costs within four weeks. The contract is silent on the impact of the *Project Manager* being late with this action; this could easily lead to a dispute. If the *Project Manager* assessed the amount, say, six weeks after the *condition* was met, does this mean that nothing is now due? Is interest due on the delayed period? In situations where the contract is silent, we inevitably see imaginative interpretations by both Parties as to what the delayed assessment means. Far better that the assessment is undertaken by the *Project Manager* on time and, if necessary, with input from the *Contractor*.

Gerrard and Waterhouse
ISBN 978-0-7277-6404-1
https://doi.org/10.1680/necrad.64041.039
ICE Publishing: All rights reserved

Chapter 5
Payments

37. Introduction to payments

In a book about disputes, payments will never be far from the discussion. Almost all, if not all, disputes will boil down to money eventually. Many disputes are about money, for example

- *Contractors* seeking payment for works completed, typically varied works
- *Contractors* seeking additional time to complete their obligations (this inevitably becomes about money)
- arguments about Defects and the quality of work, which will cost money to rectify (How much, and who should pay?)
- disagreements about ancillary documents, such as bonds, guarantees, warranties etc.

Payments in ECC contracts can be a difficult subject. For a start, there are several places in the contract where the assessment and payment are described.

1. The core clauses, primarily Section 5.
2. The main Option selected, A, B, C, D or E.
3. Many secondary Option X clauses have a financial effect, for example, Clauses X1, X6, X7, X14, X16, X17 and X20, although it could be argued that they all have an impact on payment.
4. Secondary Option Y clauses. NEC published Clause Y(UK)2 to ensure the ECC's compliance with construction payments legislation in Great Britain.[i] Similar legislation exists in other jurisdictions; Parties should obtain local advice to ensure that their contracts are compliant.
5. Both parts of the Contract Data will contain information that is necessary to assess payments, for example, *fee percentage, currency of the contract, interest rate, value engineering percentage* etc.
6. The Schedule of Cost Components or the Short Schedule of Cost Components.
7. Secondary Option Z clauses, agreed by the Parties, can and do change all the earlier items in this list.

The inclusion of six main Options in the ECC provides six different ways of assessing the payment due to the *Contractor* and consequently six different ways of allocating risk between the Parties. The choice of main Option has a considerable impact on the processes of assessing and processing payments. Essentially, although there are six different ways of assessing payments in the ECC, they can be segregated into two.

1. Options A and B are priced contracts. The *Contractor* gets paid the agreed Prices for completing the *work*s as defined in the contract. Compensation events are assessed on the basis of Defined Cost and the Short Schedule of Cost Components.
2. Options C, D, E and F are differing forms of cost reimbursement contracts, which require the *Client* to reimburse the *Contractor's* actual costs, provided they are incurred in accordance with the contract. The same process of assessment applies to compensation events as it does to the *work*s identified in the Scope at the Contract Date.

Unsurprisingly, we tend to see higher incidences of disputes among contracts using main Options C, D, E and F. The additional obligations placed on both *Contractor* and *Project Manager* provide opportunities for the effects of subjectivity to creep into assessments and consequently affect relationships. In contracts formed using main Options A and B, the Prices have been established through competitive tender or through negotiation prior to the Contract Date and therefore there is nothing to debate or disagree, at least in outline.

38. Payment provisions in the core clauses

The core clauses start the payment provisions in the ECC and appear deceptively simple.

39. The application and assessment processes

Clause 50.1 requires the *Project Manager* to assess the amount due at each assessment date. This is a clear obligation. Read it in conjunction with Clause 10.1, where the *Project Manager* is required to act as stated in the contract. So, the *Project Manager* must perform this task in accordance with the contract and certainly not to assist the commercial objectives of the *Client*. Allegations of *Project Manager* bias are frequent but rarely proven; however, failing to be impartial is a sure way for the *Project Manager* to create a dispute. In addition to the contract's clear requirements for fairness, case law (at least in England and Wales) has also intervened, with the 1974 case of *Sutcliffe* v *Thackrah*, which established that contract administrators must act in a fair and unbiased manner.[ii] A more recent case in 2005 involved an NEC2 contract, where the judge passed comments (albeit *obiter dictum*) extending that principle to the *Project Manager* in the ECC.[iii] So, the ability of the *Project Manager* to both carry out this task properly and command the respect of the *Contractor* in doing so are key to maintaining a dispute-free relationship.

The second part of Clause 50.1 requires the *Project Manager* to do something; its importance is often missed.

> Clause 50.1
>
> […] The first assessment date is decided by the *Project Manager* to suit the procedures of the Parties and is not later than the *assessment interval* after the *starting date*. Later assessment dates occur at the end of each *assessment interval* […]

Such a simple requirement for the *Project Manager*, isn't it? Almost all contracts identify an assessment interval of a month, which means that the assessment date will always be the *XX*th of the month, once the *Project Manager* has decided that the first assessment date will be the *XX*th.

While the contract requires the *Project Manager* to make this decision, it does not require anything further, such as the notification of this date to the *Contractor*; we think that this is an omission. We often see this important decision omitted by *Project Managers*, leaving some doubt about the date when assessments should commence each month. This has a major impact in 'smash-and-grab' adjudications, which are increasingly common. A smash-and-grab adjudication is one where a *Contractor* takes advantage of either the failure of the *Project Manager* to issue a certificate on time or the failure of the *Client* to issue a 'pay less' notice on time. Having failed to issue the requisite documents, the *Client* may find itself having to pay whatever sum the *Contractor* had applied for. But for a case to succeed in adjudication, the *Contractor* must be able to demonstrate that the relevant document was late; to do that, the timings refer to the assessment date, which, in turn, is linked to the first assessment date. Such omissions are less likely if the required dates are established beyond debate. Therefore, our advice is that the *Project Manager* should decide the assessment date and communicate this to both the *Contractor* and the *Client*.

Clause 50.1 goes on to say that 'later assessments' occur until the issue of either the Defects Certificate or a termination certificate. This suggests that the issue of either of these two certificates will lead to the final assessment, but the wording does not quite state that. It is rare, but not impossible, for one of the Parties to make a claim against the other for a payment assessment error some years after the Defects Certificate has been issued. The words of the contract suggest that a Party doing this would be too late to succeed, but there may be wider legal arguments if it could be shown that a payment was otherwise due. The final assessment provisions, described later in this chapter, perhaps close out this opportunity if they are utilised correctly.

The *Contractor* is required to make an application for payment and submit it to the *Project Manager*. Clauses 50.3 and 50.4 prohibit any payment to the *Contractor* in the absence of an application for payment. The application should

- be submitted before each assessment date
- set out the amount the *Contractor* considers to be due
- include details of how that amount has been calculated
- be in the form set out in the Scope.

The *Contractor* should comply with these four requirements to ensure that the *Project Manager* is able to undertake its assessment and certification duties on time and with a greater chance of agreeing with the *Contractor's* own assessment. The *Project Manager* only has a week to certify the payment; this can be a challenge with contracts using main Options C, D, E or F. Therefore, the *Project Manager* might not have time to go looking for information that should have been kept by the *Contractor*. The contract does not require the *Project Manager* to go hunting for information when making assessments, so the *Contractor* should make sure that everything necessary is made available, and that the application is submitted on time. Disagreements (as distinct from disputes) frequently occur over the amount due; only communication can reduce the incidences of these. Where such disagreements develop into disputes, the *Adjudicator* will only make a decision based on the documents provided in the submissions.

40. The Price for Work Done to Date

Clause 50.3 sets out the amount due, if an application has been submitted by the *Contractor*.

> **Clause 50.3**
>
> [...] the amount due at the assessment date is
>
> - the Price for Work Done to Date,
> - plus other amounts to be paid to the *Contractor*,
> - less amounts to be paid by or retained from the *Contractor*.

Other amounts paid to the *Contractor* could include

- interest for late payment (Clause 51.4)
- insurance costs if the *Client* does not provide insurance (Clause 86.3)
- price adjustment for inflation (Clause X1)
- bonus for early Completion (Clause X6)
- Key Performance Indicators (Clause X12 or X20)
- advanced payment to the *Contractor* (Clause X14.1).

Amounts to be paid by or retained from the *Contractor* could include

- *Client's* additional costs resulting from a failure to achieve a *condition* by the identified Key Date (Clause 25.3)
- failure to submit the first programme on time (Clause 50.5)
- insurance costs if the *Contractor* does not provide insurance (Clause 85.1)
- delay damages (Clause X7)
- recovery of advanced payment (Clause X14.3)
- retention (Clause X16)
- low performance damages (Clause X17).

Wherever additional amounts are claimed by the *Contractor* or where the *Project Manager* and *Client* wish to deduct payments from the *Contractor*, the relevant person should clearly set out the entitlement and its quantum, in accordance with Clauses 50.2 and 51.1, respectively. Again, clear communication at this stage will help to avoid any dispute about the assessment.

The primary component of the assessment is the Price for Work Done to Date. The smattering of capital letters throughout this phrase indicates that it is a defined term. If you're working from the 'black book'[iv] of the ECC publications, then you won't find the definition on pages 3 and 4, as you do with most defined terms. Each main Option has its own definition of this term, reflecting the differing risk allocations of the six Options. We explain what each one means later in this chapter, but the definitions are found in the main Option clauses in the black book. Each definition points to other areas of the contract, for example an Activity Schedule, a Bill of Quantities or a Schedule of Cost Components etc. The green books provide all of the defined terms for the core clauses and main Options in the same two or three pages, making for simpler reading.

Clause 50.6 often has a role in payment disputes.

> **Clause 50.6**
>
> The *Project Manager* corrects any incorrectly assessed amount due in a later payment certificate.

Both Parties, therefore, can 'revisit' previous assessments, although the revisiting needs to comply with the provisions of Clauses 50.1 and 53. Clauses 51.2, 51.3 and 51.4 require the payment of interest in several situations, including '*in relation to a mistake*'. Where the *Project Manager* 'corrects' a previous assessment after the submission, by the *Contractor*, of more information, we suggest that this did not result from a mistake and therefore interest is not due. This situation is common, particularly where the *Contractor* produces accounts and records to substantiate a cost many months after it was incurred.

The charging for, and payment of, interest can have a puzzling effect on relationships. A Party that has knowingly withheld hundreds of thousands of pounds in breach of contract suddenly gets very annoyed when the other Party rightly demands hundreds of pounds in payment of interest that is also contractually due. Disputes involving money (which is nearly all of them) typically have a claim for interest attached to them, but often in situations where the receiving Party has not previously applied for it. The *Project Manager* has a duty to assess the amount due in accordance with the contract and therefore must include interest where it is due. But underpayments by the *Client* are rarely straightforward and the majority that reach a dispute will result from a decision of the *Project Manager*, most commonly the assessment of compensation events. *Project Managers* who see their role as minimising the amount payable by the *Client* should consider what happens when that incorrect conduct leads to disputes.

The Contract Data contains an entry for an '*interest rate*', which it tells us should be not less than 2% above the stated bank base rate. This specification of a minimum of 2% is an oddity; it is surely for the Parties to decide what is appropriate. Why 2%? Why not 10%? Why not 0.5%? In short, there is little to justify the minimum stated in the published form.

The UK[v] has legislation that implies a term into contracts to require the payment of interest on contractual debts (but not damages) of 8% above base rate if the contractual parties have not included an adequate remedy for the effect of late payment in the contract. A common claim in adjudication is that this 'statutory' rate of 8% above base should apply even if the Parties have agreed a different (i.e. lower) rate in the contract. The claiming party typically states that despite having agreed, say, 3% above base rate, that is not an adequate remedy and therefore the (higher) statutory rate should apply instead. Case law has yet to provide us with a substantive answer to this issue. Users of the ECC in the UK should note that there is no link between the statutory requirement for an adequate remedy and the minimum rate suggested by the ECC Contract Data.

It is difficult to avoid some of the issues that arise from the payment of interest, but the basics of good contract administration should apply.

- Ensure that the Contract Data is properly completed with the *interest rate* and the name of an appropriate bank which publishes a rate and the title of that rate.
- If the interest rate chosen is less than 2%, remove the comment '*not less than 2*' from the document.
- The *Contractor* should apply for interest where it is contractually due in payment applications.
- The *Project Manager* should assess and certify interest in accordance with the contract.

41. Payment processes in the core clauses

The correct assessment of payments in the ECC requires participants to read several sections of the contract; this requires some skill on behalf of both the *Project Manager* and the *Contractor*. It is easy to put a step wrong when looking at five or six different parts of the contract simultaneously. Therefore, close collaboration between the *Project Manager* and the *Contractor* is essential; each will probably have something to learn from the other.

For those in a jurisdiction with construction payments legislation, the provisions of Clause 51 will have to be read in conjunction with additional terms of contract to accommodate the requirements of that legislation. Within the UK those additional terms are contained within Clause Y(UK)2, which we describe later. Our description of Clause 51 is of its unamended status, i.e. with no additional clause included for legislative reasons.

Clause 51.1 is straightforward.

- The *Project Manager* certifies a payment within one week of the payment date.
- The certificate must include details of how the amount has been assessed.
- Payments can be from *Client* to *Contractor* or vice versa. Payments made by the *Contractor* to the *Client* are possible, if rare. The typical scenario is towards the end of a project, where the application of delay damages is greater than the value of the work completed.

Our advice is similarly straightforward; the *Project Manager* should complete this task on time (remembering our earlier advice about the need to set the first assessment date) and provide enough explanation to the *Contractor* about how the amount has been calculated. The need for effective communication is paramount; these exchanges should not be merely documentary. Monthly certifications should involve face-to-face discussion, particularly with main Options C, D, E and F.

Payment must be made by the paying Party (remembering that in some circumstances this may be the *Contractor*) within three weeks of the assessment date. There is a provision in the Contract Data for the *Client* to identify a different payment period if three weeks is not appropriate. Many disputes have their origins in late payments even before the quantum of the payments is debated. In terms of managing constructive relationships, it is preferable that the *Client* specifies a long payment period (say five weeks) and then honours it repeatedly, rather than specifying a shorter period and repeatedly paying late. The predictability of receipts is important to most companies in the management of cash flow.

42. Defined Cost

Defined Cost plays a major role in payments in the ECC, primarily in contracts using main Options C, D, E or F. The only role for Defined Cost in contracts using main Options A or B is in the assessment of compensation events. In main Options C, D, E and F, Defined Cost has a much more involved role in payments. The term 'Defined Cost' is defined in each of the main Options and we explain those definitions later in this chapter.

Defined Cost is paid by the *Client* to the *Contractor* in situations where the *Client* is obliged to reimburse the *Contractor* the costs incurred either in the short term or permanently

- **main Options A and B:** used in the assessment of compensation events
- **main Options C, D, E and F:** used in the assessment of all *works*.

Clause 52.1 sets out some useful ground rules for Defined Cost that apply in all situations.

1. Any of the *Contractor's* costs not included as Defined Cost are included in the Fee. The Fee, defined in Clause 11.2(10) is 'the amount calculated by applying the *fee percentage* to the amount of Defined Cost.' The *fee percentage* is identified in Contract Data part two by the *Contractor*. The Fee provides for the recovery of all costs that are not included in the contract's definition of Defined Cost. Typically, these are costs away from the Working Areas, such as head office costs or costs incurred within the *working areas* that cannot be charged to the *Client*.
2. Defined Cost amounts should follow these principles.
    I. They should use rates and percentages included in the Contract Data (i.e. those agreed by the Parties, usually during a competitive process). The use of these should not cause too much concern as the numbers concerned should be a matter of fact.
    II. They should use other amounts at open market or competitively tendered prices. In other words, the *Client* will only have to pay the 'going rate' for an activity if its price is not included in the contract already. Proving that an amount is at 'open market' rates can be a challenge, particularly where a *Contractor* is purchasing an item or a service from a related company within the same group; disagreements about these issues are common and can lead to disputes. If the *Client* has requirements for how a part of the *works* should be procured and wishes it to be competitively tendered, the Scope should include a procurement process for the *Contractor* to follow and the *Contractor* should ensure that the process is adhered to.
    III. They should include deductions for all discounts, rebates and taxes that can be recovered. These requirements can prove quite tricky in practice. The contract requires that the *Client* should benefit from any benefits that the *Contractor* can obtain. A typical supply arrangement between a *Contractor* and a supplier might agree an end-of-year discount based on the volume of business transacted between the two of them. It is therefore difficult for a *Project Manager* to know what the net price of an item is midyear. In this situation, a reasonable estimate could be agreed between the *Contractor* and the *Project Manager*. If the discount is sufficiently large, then it could be revisited at the year-end for the supply deal. Our advice here is to
        i. base any interim estimates on whatever documentary evidence is available
        ii. keep good records of what has been assessed and why
        iii. communicate closely.

43. The Fee

The Fee is defined in Clause 11.2(10) and is calculated using the *fee percentage* identified in Contract Data part two by the *Contractor*, most probably as a part of its tender response during procurement. The calculation of the Fee is unlikely to lead to any disagreement or debate between the Parties as it is so simple; applying the *fee percentage* to the Defined Cost. Provided the *fee percentage* is clearly and correctly identified in the Contract Data, we see no reason for further commentary.

44. Final assessments

The introduction of final assessments in NEC contracts with the publication of NEC4 is to be welcomed in terms of dispute avoidance. Without final assessments, in earlier versions of NEC contracts, the loose wording of Clause 50.1 mentioned earlier always left open the opportunity for either Party to recommence contractual hostilities some time after Completion. In the UK, a right to adjudication 'at any time'[vi] has encouraged some Parties to seek a reassessment of amounts due many months or years after Completion, not always successfully. By providing the Parties with the ability to finalise assessments, the contract enables both to draw a line under part or all of the project and gain certainty over those payments. Certainty always assists in avoiding disputes.

Final assessments work in two ways.

1. For the assessment of the final amount due and the consequent certification of the final payment. These provisions are in Clauses 53.1 to 53.4 and apply to all main Options.
2. For the finalisation of parts of Defined Cost when using main Options C, D, E and F. These provisions are in Clause 50.9.

We look at the final assessments in main Options C, D, E and F later in this chapter but, here, we describe what happens at the end of the project.

The process starts with the *Project Manager* assessing the final amount due and issuing a certificate no later than

- four weeks after the *Supervisor* has issued the Defects Certificate, or
- 13 weeks after the *Project Manager* has issued a termination certificate.

Note that the time limit is linked to when the relevant certificate is issued, rather than when it should be issued. So, the *Supervisor* or the *Project Manager*, as appropriate, should ensure that the relevant certificate is issued on time. Any delay in doing so will have an impact on the final assessment process and potentially delay the payment of money due. *Clients* should also bear this process in mind when setting the *defects date*. We occasionally see contracts setting the *defects date* as late as six years after Completion. While there may be good reasons for doing this in terms of dealing with Defects, the *Client* should bear in mind what other processes will be affected by such decisions. Responsibility for issuing the Defects Certificate should be remembered; *Supervisors* are often dispensed with at Completion. In that situation, their duties should be allocated elsewhere. Similarly, when dealing with termination, almost always a breeding ground for disputes, the *Project Manager* should issue the termination certificate punctually before then dealing with the payment aspects of termination.

If the *Project Manager* does not make the assessment within 'the time allowed', the *Contractor* can make its own assessment and issue it to the *Client* (note, not the *Project Manager*). There is no payment certificate in this situation where the assessment is made by the *Contractor*.

Whoever produces the final assessment must provide details of how the final amount has been calculated.

There are a few issues here to assist with avoiding disputes.

- Ensure that someone remains appointed as *Supervisor* and is in position at the *defects date*.
- Ensure that the *Supervisor* issues the Defects Certificate on time.
- The *Project Manager* should issue a termination certificate on time, where one is needed. We describe the termination provisions later in this book but there are other timed obligations on the *Project Manager* there too.

- The *Contractor* should issue its own final assessment if the *Project Manager* does not.
- The *Client* should remain alert to the timing of these actions, particularly where the *Project Manager* is remiss in undertaking the required duties.
- Close communication between the *Contractor*, *Project Manager* and *Client* will be needed, particularly with the volume of information that will be exchanged, some of which will require detailed explanation.

Of course, the finalisation of anything under contract can and does lead to disputes. Clause 53.3 recognises this possibility and provides a route for resolving such a dispute. In fact, three routes are provided, one for each of Options W1, W2 and W3 for resolving and avoiding disputes. Only one of these Options can apply to each contract; this is identified in Contract Data part one. The three routes for dealing with disputes on final assessments reflect the differing nature of the three dispute resolution Options. Let's look at Clause 53.3.

> Clause 53.3
>
> An assessment of the final amount due issued within the time stated in the contract is conclusive evidence of the final amount due under or in connection with the contract unless a Party takes the following actions [...]

This clause is wide-ranging. Look at the statements 'under or in connection with' and 'conclusive evidence'; these really restrict the ability of a Party to challenge a final assessment after the conclusion of the contractual processes.

There are three routes to resolving a dispute about a final assessment.

1. Where Option W1 is used, the referring Party refers the dispute
    I. first, to the *Senior Representatives* within four weeks of the assessment
    II. then, for any matters not resolved by the *Senior Representatives*, to the *Adjudicator* within three weeks of the *Senior Representatives* producing a list of the issues not agreed
    III. finally, to the *tribunal* within four weeks of the *Adjudicator's* decision being made.
2. Where Option W2 is used, the referring Party refers the dispute
    I. first, to the *Senior Representatives* or to the *Adjudicator* within four weeks of the assessment (this choice ensures that the contract complies with the statutory right to adjudication at any time)
    II. then, for any matters not resolved by the *Senior Representatives*, to the *Adjudicator* within three weeks of the *Senior Representatives* producing a list of the issues not agreed
    III. finally, to the *tribunal* within four weeks of the *Adjudicator's* decision being made.
3. Where Option W3 is used, the referring Party refers the dispute
    I. first, to the Dispute Avoidance Board
    II. then, and finally, to the *tribunal* within four weeks of the Dispute Avoidance Board's recommendation being made.

These processes have some common themes; once triggered, there are time bars for the referring Party to adhere to. The processes must be read in conjunction with the other drafting in Options W1, W2 or W3. For example, the reference to the *Adjudicator* in Options W1 or W2 then sees the relevant process take over at that point. All three processes seek to resolve the dispute through informal negotiation first; that is the role of the *Senior Representatives* and the Dispute Avoidance Board. But, clearly, where that fails there needs to be a formal resolution process that leads to a binding decision, even if not agreement.

Clause 53.4 allows the amendment of the assessment of a final amount due

1. after an agreement of the Parties: we advise that this is recorded in writing as a change to the contract, in compliance with Clause 12.3
2. after a decision by the *Adjudicator* or a recommendation of the Dispute Avoidance Board, where that decision or recommendation has not been referred to the *tribunal* within four weeks. In this situation, it will be worthwhile for one Party, or possibly both, to notify the other that the amount due is now considered to be final following the other Party's failure to issue a notice of dissatisfaction. Clear communication at this stage will prevent costly debate later.

45. Clause Y(UK)2

Several jurisdictions have introduced legislation providing statutory intervention in the otherwise freedom of parties to enter into contracts on terms of their own choosing. While the NEC contracts are drafted for use anywhere, their dominant position in the UK construction industry has led to UK-specific drafting being included in the contracts.

The relevant legislation in the UK is

- the Housing Grants, Construction and Regeneration Act 1996, as amended by the Local Democracy, Economic Development and Construction Act 2009, which applies to England, Wales and Scotland
- the Construction Contracts (Northern Ireland) Order 1997, as amended by the Construction Contracts (Amendment) Act (Northern Ireland) 2011, which applies in Northern Ireland.

Both pieces of legislation have, between them, spawned four statutory instruments, known as 'Schemes'. There is a Scheme for each territory of Northern Ireland, Scotland, Wales and England. The details of the Schemes fall outside the subject matter of this book.

Options W1 and W2 accommodate the dispute resolution provisions of the legislation; we describe this in Chapters 11 and 12. Option Y(UK)2 accommodates the payment provisions of the legislation.

The legislation requires construction contracts[vii] to contain provisions for

1. a due date for payment (provided in Clause Y2.2)
2. a final date for payment (Clause Y2.2)
3. payment notices (Clause Y2.2)
4. 'pay less' notices (Clause Y2.3)
5. the consequences of the *Contractor* suspending performance in the event of non-payment by the *Client* (Clause Y2.5).

The inclusion of Clause Y(UK)2 is something that the Parties typically cannot avoid. While it dovetails with the other payment provisions of the contract, it does place additional obligations on the Parties, which must be adhered to. These are primarily on the *Client* and *Project Manager*, who must remain alert to the possibility of a smash-and-grab adjudication.

Our advice here is simple.

- Issue the required notices on time.
- Ensure that the notices contain the requisite information.

Where this advice is followed, the possibility of a dispute over the failure to administer the contract properly is reduced.

46. The main Option provisions

The ECC has six main Option clauses, each designed by the drafters to provide a particular allocation of risk between the Parties. The contract is designed to use only one main Option. Occasionally, we see contracts formed where the Parties (usually prompted by the *Client*) have agreed terms that include more than one main Option. This is not to be recommended. In that situation, inevitably, even with the addition of some judiciously-drafted Z clauses, situations will arise where the assessment of a payment has more than one possible answer and the Parties' own views will follow the money. Our description of the six main Options follows and is based solely on the use of one main Option per contract.

47. Main Option A – priced contract with Activity Schedule

Option A uses an Activity Schedule and we explain its use here. Much of what we say here also applies to the use of an Activity Schedule in Option C.

Option A is probably the simplest of the six payment mechanisms to deal with. The Activity Schedule lists activities that, once completed, lead to the certification of an agreed payment at the next certification. The Activity Schedule does not need to list every single activity in the works and it is probably best that it does not. A few contractual provisions are worth stating here.

> Clause 11.2(21)
>
> The Activity Schedule is the *activity schedule* unless later changed in accordance with these *conditions of contract*.
>
> Clause 11.2(29)
>
> The Price for Work Done to Date is the total of the Prices for
> - each group of completed activities and
> - each completed activity which is not in a group.
>
> A completed activity is one without notified Defects the correction of which will delay following work.

> Clause 11.2(32)
>
> The Prices are the lump sum prices for each of the activities on the Activity Schedule unless later changed in accordance with the contract.
>
> Clause 31.4
>
> The *Contractor* provides information which shows how each activity on the Activity Schedule relates to the operations on each programme submitted for acceptance.
>
> Clause 55.1
>
> Information in the Activity Schedule is not Scope or Site Information.

The contract goes on to say that where there is not a suitable relation between the Activity Schedule and the Scope or the Accepted Programme, the Activity Schedule should be changed by the *Contractor* and submitted to the *Project Manager* for acceptance. As with other acceptance tests in the contract, the reasons why the *Project Manager* can decline to accept the revised Activity Schedule are listed.

Clause 11.2(21) reflects the likelihood that the Activity Schedule will change during the *works*. On the Contract Date, the Activity Schedule is the *activity schedule*, i.e. the one identified in the Contract Data part two. But, with changes (for example, compensation events, accepted Defects, acceleration, changes in methods), that will be superseded in due course.

It is important for reasons of contractual interpretation to recognise that the Activity Schedule is a stand-alone document. It is identified in Contract Data part two but is not a part of it. Clause 55.1 explains that the Activity Schedule is neither Scope nor Site Information, so exempting the schedule from provisions such as, for example, Clauses 14.3, 20.1 and 60.3.

As with much other dispute avoidance advice, we start by recommending that the schedule is drafted carefully so that interpretation problems do not arise. The definition of Price for Work Done to Date shown previously demonstrates the importance of understanding what a complete activity means. The *Project Manager* is required to assess and certify the payments associated with completed activities. An activity that is 100% complete will lead to payment in the next certificate; one that is 99% complete will not. There is no partial certification allowed, for example to pay 75% of the amount stated. The process is designed to be straightforward to operate and to minimise the administrative burden of both Parties.

This means that the drafting of words in the Activity Schedule must be precise. Here are some examples.

Example 5.1

| Undertake Site clearance works and establish office & welfare facilities | £40,000.00 |

Example 5.2

1002	Erect Site temporary fencing as per drawing XXX-123	£5,000.00
1003	Strip and store topsoil as per drawing XXX-128	£8,000.00
1004	Remove and dispose waste material from Site as required in Scope para. S207	£9,000.00
1005	Erect Site office and welfare facilities as per drawing CCC-327	£18,000.00

The first example leaves much to be interpreted and therefore means that there may be a disagreement between the *Contractor* and the *Project Manager*.

Example 5.2, in contrast to Example 5.1, benefits from

- greater granularity in explaining activities, therefore cash flow should improve
- numbering of activities will replicate that in the Accepted Programme, thus complying with Clause 31.4
- linking activities to items in the Scope, so there can be no debate about whether the activity has been completed in accordance with the contract.

The requirement to complete activities fully to justify payment means that some disputes inevitably occur when the *Project Manager* does not certify payments in respect of activities that are nearly complete.

The Activity Schedule does not mention quantities; it is up to the *Contractor* to assess these at tender stage and decide how it wishes to complete the Activity Schedule. Regarding the drafting of Activity Schedules, we recommend minimal specification or requirements from clients in tender documents; the *Contractor* should be granted maximum flexibility in preparing its response.

When disputes do occur, each Party will pore over the wording and seek to demonstrate an interpretation that suits its own argument. Most disputes about the interpretation of Activity Schedules do not develop, as the argument is typically about <u>when</u> something is paid, as opposed to <u>how much</u> is paid. But good drafting at the tender stage and close adherence to it during construction should minimise any problems. We see very few disputes involving Activity Schedules, except where compensation events are involved.

Defined Cost only has a role in main Option A contracts when compensation events are assessed; we describe this in Chapter 6.

48. Main Option B – priced contract with Bill of Quantities

Option B uses a Bill of Quantities; we explain its use here. Much of what we say here also applies to the use of a Bill of Quantities in Option D.

Option B is the closest to what we might term a 'traditional' contract, relying on a Bill of Quantities for its assessment. Here are the relevant clauses.

Clause 11.2(22)

The Bill of Quantities is the *bill of quantities* unless changed in accordance with these *conditions of contract*.

Clause 11.2(30)

The Price for Work Done to Date is the total of

- the quantity of the work which the *Contractor* has completed for each item in the Bill of Quantities multiplied by the rate and
- a proportion of each lump sum which is the proportion of the work covered by the item which the *Contractor* has completed.

Completed work is work which is without notified Defects the correction of which will delay following work.

Clause 11.2(33)

The Prices are the lump sums and the amounts obtained by multiplying the rates by the quantities for the items in the Bill of Quantities.

An important entry is also needed in Contract Data part one, to identify the *method of measurement*. There is a skill to working with Bills of Quantities that is not as prevalent now as it once was, owing to the declining popularity of bills as a method of measuring work. Methods of measurement are published by various industry bodies and aid the preparation and use of Bills of Quantities. The contract requires the use of the selected *method of measurement* to aid the interpretation of the Bill of Quantities. Without it, problems will inevitably result.

Currently in the UK, the available methods of measurement are

- the Civil Engineering Standard Method of Measurement[viii]
- the Method of Measurement for Highway Works, which is part of the *Manual of Contract Documents for Highway Works*[ix]
- the Rail Method of Measurement
- the Standard Method of Measurement
- the New Rules of Measurement, published by the Royal Institute of Chartered Surveyors.

Our advice is to do the following.

- The *Client* should select a *method of measurement* carefully, ensuring that it is appropriate to the *works*.
- The *method of measurement* should be clearly identified in Contract Data part one, including all version or edition numbers and references.
- The *Client* should prepare the Bill of Quantities in full accordance with the *method of measurement*.
- Tenderers (one of whom will become the *Contractor*) should prepare their tenders in accordance with the *method of measurement*.
- The *Contractor* should prepare payment applications in compliance with the Bill of Quantities and the *method of measurement*.
- The *Project Manager* should assess and certify amounts due in compliance with the *method of measurement*.

All of this requires knowledge of the process; therefore, only those people with the requisite skills should be used to undertake these tasks.

Clause 11.2(22) reflects the likelihood that the Bill of Quantities will change during the *works*. On the Contract Date, the Bill of Quantities is the *bill of quantities*, i.e. the one identified in the Contract Data part two. But, with changes (for example, compensation events, accepted Defects, acceleration or significant changes in quantities) that will be superseded in due course.

It is important, for reasons of contractual interpretation, to recognise that the Bill of Quantities is a stand-alone document. It is identified in Contract Data part two but is not a part of it. Clause 56.1 explains that the Bill of Quantities is neither the Scope nor Site Information, so exempting the Bill of Quantities from provisions such as, for example, Clauses 14.3, 20.1 and 60.3.

The ECC has one 'precedence' clause when working with Bills of Quantities. Look at Clause 60.7.

> Clause 60.7
>
> In assessing a compensation event which results from a correction of an inconsistency between the Bill of Quantities and another document, the *Contractor* is assumed to have taken the Bill of Quantities as correct.

So, in the event of any inconsistency, the Bill of Quantities takes precedence; another reason for ensuring that the Bill of Quantities is properly prepared.

The assessment of the Price for Work Done to Date under this Option relies on the physical measurement of the *works*. This again requires properly skilled people and, crucially, time. The *Project Manager* and the *Contractor* should consider joint measurement to reduce the cost and time of the exercise and (hopefully) reduce the possibility of disputes.

Where disputes do occur in the measurement of work under Bill-of-Quantities contracts, an agreed factual record of what occurred will assist those wanting to resolve the dispute. Where the two Parties have conflicting measurements, that task is complicated significantly.

49. Assessment of Defined Cost under main Options C, D and E

All three of these Options share a common definition of the Price for Work Done to Date.

> Clause 11.2(31)
>
> The Price for Work Done to Date is the total Defined Cost which the *Project Manager* forecasts will have been paid by the *Contractor* before the next assessment date plus the Fee.
>
> Clause 11.2(24)
>
> Defined Cost is the cost of the components in the Schedule of Cost Components less Disallowed Cost.

While these three Options share these two common definitions, they differ in their use of the term 'Prices'; we describe that impact later in this chapter.

If we look at the definition of the Price for Work Done to Date, we can see that the *Project Manager* needs to make a judgement of what will be paid in the following month, not what will be accrued or incurred (emphasis added). To some extent this should be straightforward. Amounts being paid to Subcontractors and suppliers in the following month will almost certainly be the subject of invoices that have already been received, so immediately there is some evidence of the obligation for payment. The *Project Manager* will then need to establish whether the amount should be paid based on whether it was expended to Provide the Works in accordance with the Scope. Forecasts of currently unascertained costs, for example people, can be made and documented and then compared with actual figures a month later. The *Project Manager* will never get the forecast entirely the same as the subsequent reality; hence, Clause 50.6 is important for correcting amounts in future assessments.

We explained the rules of Defined Cost earlier in this chapter. Disputes about the assessment of Defined Cost tend to focus on one or both of these areas.

- Was the amount incurred at all, or at the quantum claimed?
- Should the amount be reimbursed in accordance with the contract?

Solving the first area should, in theory, be relatively straightforward. Most transactions create some type of documentation and therefore an audit route too. But in reality, an individual project will possibly create millions of such transactions and, in the *Contractor's* wider business, many millions more. Two examples from Option C contracts follow.

Example 5.3

A construction manager, an employee of the *Contractor*, identified that several key components for a critical activity had not been ordered. The components were needed for the same day. She drove to a local builders' merchant and paid for them on her company charge card. The receipt got thrust into her jacket pocket and forgotten about.

Somewhat disorganised at claiming her expenses, she only submitted her claim six months later, reconciling the payment with the entry on her charge card statement. The receipt from the builders' merchant was, by now, in a sad condition and didn't state any helpful information. The receipt quoted a stock number, rather than the name and the type of product. So, we know that she purchased 80 of CDDEEF18765443/A, whatever they are! She fortunately remembered what they were and could explain.

But the *Project Manager* wanted documentary evidence. That required the *Contractor* to go back to the builders' merchant and get- a statement about the actual products purchased.

All of this revolved around £300 worth of stainless steel bolts. But the tardy nature of the construction manager's administration and the consequent six months delay created far more work for the project than was necessary. It is common to see protracted debates about relatively low-value transactions. Those involved in such debates should consider the 80/20 rule.

Example 5.4

In a second example, an English *Contractor* has been working in Scotland. The displaced geographical knowledge of several people involved in this process didn't help. The project involved the demolition of a disused industrial plant. Contamination of the demolition materials was feared and therefore frequent samples were taken and transported to a laboratory in England for testing.

The Site was near the Scottish town of Dunbar, not somewhere known to the accounts manager at the laboratory. She mistakenly labelled invoices with the name 'Dundee'. A Scottish city with a similar name, albeit a 100-miles drive away.

Over £40,000 of testing was undertaken. Each certificate and each invoice was labelled 'Dundee project'. The invoices for this work were issued electronically to the generic accounts payable email address at the *Contractor*'s head office. In fact, the *Contractor* has outsourced its accounts payable function to a processor in Bangalore, India. The people there, it might be assumed, knew just as much about the geography of Scotland as the people at the English test laboratory.

The invoices were incorrectly coded in Bangalore and were allocated to the wrong project. By the time that the invoice arrived on the accounts of the correct project, eight months had passed. The *Project Manager* then disallowed the costs in their first submission because of the failure to identify the correct name of the project. The passage of time meant that the *Project Manager* was less inclined to be flexible than might otherwise be the case. £40,000 was now disputed. No doubt, further investigation and some flexibility will resolve this issue, but all this additional work would not be necessary had there been better discipline and checking at the time.

What do we learn from these examples? Both involve relatively small amounts of money, but the reimbursement of both has proved more difficult than it otherwise might have been, owing to the poor record keeping of the *Contractor* and the inflexibility of the *Project Manager*.

Establishing whether something should be reimbursed in accordance with the contract is often more difficult than deciding the quantum of that amount. The contract contains many rules prescribing what is and what is not due for reimbursement; hence the amount of space we dedicated to the subject earlier in this chapter.

The definition of Disallowed Cost is lengthy, and it is necessary for us to repeat it here as its role in disputes is common. While it enjoys its own definition, it is also a key component of the definition of Defined Cost. This definition, in Clause 11.2(26), applies to all three main Options.

> Clause 11.2(26)
>
> Disallowed Cost is cost which
>
> - is not justified by the *Contractor's* accounts and records,
> - should not have been paid to a Subcontractor or supplier in accordance with its contract,
> - was incurred only because the *Contractor* did not
>   - follow an acceptance or procurement procedure stated in the Scope,
>   - give an early warning which the contract required it to give or
>   - give notification to the *Project Manager* of the preparation for and conduct of an adjudication or proceedings of a tribunal between the *Contractor* and a Subcontractor or supplier
>
> and the cost of
>
> - correcting Defects after Completion,
> - correcting Defects caused by the Contractor not complying with a constraint on how it is to Provide the Works as stated in the Scope,
> - Plant and Materials not used to Provide the Works (after allowing for reasonable wastage) unless resulting from a change to the Scope,
> - resources not used to Provide the Works (after allowing for reasonable availability and utilisation) or not taken away from the Working Areas when the Project Manager requested and
> - preparation for and conduct of an adjudication, payments to a member of the Dispute Avoidance Board or proceedings of the tribunal between the Parties.

So, Disallowed Cost has many heads. The aim of having such a definition is to ensure that the *Client* is protected and is not required to pay for anything that wasn't necessary or was not properly incurred. By comparison, Defined Cost protects the *Contractor* and ensures that all properly incurred costs are reimbursed. The rules for both give the Parties the framework needed to meet these aims.

The first item is simple; the *Contractor* must be able to show from records and accounts that a payment has been made and that requires access to be given to the *Project Manager*. We discuss later the interaction between the *Contractor* and the *Project Manager*, but it is essential here, referring to Clauses 52.2 and 52.4. Where records are missing, there are usually other records that can fill the gap. But *Project Managers* need to be flexible, remembering that the standard of proof required is that of the balance of probabilities.

Incorrect payments to Subcontractors or suppliers are disallowed. The appointment of Subcontractors must involve the *Project Manager* (see Clause 26 for details). So this is the opportunity for the *Project Manager* to have an input into the prices and terms of the subcontract. Once the *Project Manager* has accepted those, it is too late to disagree with them. Payments made to a Subcontractor and a supplier, if they are in accordance with the contract in question, cannot be disallowed. The *Project Manager* is not involved with the appointment of suppliers, i.e. those providers who don't match the definition of Subcontractor in Clause 11.2(19), and therefore will need to see contracts with those organisations to make a judgement. This can prove difficult, as the requirement to provide such documents is not expressly stated in the contract. We therefore recommend that the Scope states that such documents are to be provided to the *Project Manager*, connecting with the final limb of Clause 52.2.

The contract disallows any costs that were incurred only because 'the *Contractor* did not follow an acceptance or procurement procedure stated in the Scope'. This clause provokes many disputes, owing to (i) *Clients* not putting any such procedures in the Scope or (ii) those procedures being supremely ill-equipped to the task in question. To avoid disputes here, the *Client* should provide clear and appropriate procedures in the Scope and ensure that the *Project Manager* draws the *Contractor's* attention to them before their use becomes critical. The *Contractor* should keep records that demonstrate compliance with the relevant procedure.

We described in Chapter 3 how the *Contractor's* failure to notify an early warning that was required would lead to Disallowed Cost.

There are two mentions of legal proceedings in the Disallowed Cost definition. The inference of the first reference appears to be that the cost of proceedings between the *Contractor* and a Subcontractor or supplier is allowable, provided that the *Project Manager* has been kept informed. Typically, this would be desirable in cases about the work of a Subcontractor under the subcontract that also had an impact on the main contract. By close liaison with the *Project Manager* on such an issue, the *Contractor* could possibly avoid the need for proceedings, or at least reduce their impact. There are several legal issues for the *Contractor* to consider here, such as confidentiality and common interest privilege.

The second mention of legal proceedings predictably prevents the *Client* having to pay the legal costs of the *Contractor* taking action against the *Client*. This appears to make sense to us. The final list item of Clause 11.2(26) is not well-drafted. We think that it is disallowing the cost of any adjudication between the *Contractor* and the *Client*. But the way the clause is structured, it might be read as disallowing the costs of any adjudication, including those with Subcontractors or suppliers, which appears to conflict with our earlier conclusion that such costs should be allowed.

The second list in the definition of Disallowed Cost mainly refers to costs incurred in the physical carrying out of the *works*. Keeping good records is the key once again here to demonstrating what has happened in the Working Areas and why. Once again, we have a reference to something being stated in the Scope, so the *Client* must ensure that it is included therein. We have the rare use of the word 'reasonable' in reference to wastage of Materials, a clear provocation to a disagreement in most forms of contract. People who are required to be reasonable, and that means both the *Contractor* and the *Project Manager* here, must be capable of demonstrating their reason.

Overall, the definition of Disallowed Cost provides several judgements that the *Project Manager* must make. Many can be avoided if the *Contractor* can retain and provide the requisite records. But many judgements will have to be demonstrated by the *Project Manager* where costs are disallowed.

When we consider these three definitions, we can see that the Parties must consider many variables when navigating their way through the payment procedures.

The onus is on the *Project Manager*, see Clauses 50.1 and 11.2(31). But in cost reimbursable situations, the *Project Manager* cannot perform this task without the assistance of the *Contractor*. These clauses prescribe the input of the *Contractor* to the payments process.

Clause 50.2

The *Contractor* submits an application for payment to the *Project Manager* before each assessment date setting out the amount the *Contractor* considers is due at the assessment date. The *Contractor's* application for payment includes details of how the amount has been assessed and is in the form stated in the Scope.

Clause 20.4

The *Contractor* prepares forecasts of the total Defined Cost for the whole of the *works* in consultation with the *Project Manager* and submits them to the *Project Manager* […]

Clause 26.4

The *Contractor* submits the pricing information in the proposed subcontract documents for each subcontract to the *Project Manager* […]

Clause 52.2

The *Contractor* keeps these records

- accounts of payments of Defined Cost,
- proof that the payments have been made,
- communications about and assessments of compensation events for Subcontractors and
- other records as stated in the Scope.

Clause 52.4

The *Contractor* allows the *Project Manager* to inspect at any time within working hours the accounts and records which it is required to keep.

So, while there is an onus on the *Project Manager*, there is a need for the *Contractor* and the *Project Manager* to work together here. That work should start immediately the contract is formed. The requirements of the *Project Manager* need to be matched to the vagaries of the *Contractor's* accounts and records systems. This is no mean feat, particularly in larger-value contracts. Once again, competent people are needed to do these tasks; not everyone has the financial, construction or audit knowledge necessary. The close co-operation of the *Project Manager* and the *Contractor* in respect of Clause 20.4 is vital; the efficient operation of those costs forecasts will go a long way towards providing confidence to the *Client* that these issues are being properly managed.

Parties to any type of reimbursable arrangement will typically form different views on what payments are due under that arrangement. The ECC provides a detailed structure as to what is due and does it through the following provisions.

> **Clause 52.1**
>
> All the *Contractor's* costs which are not included in the Defined Cost are treated as included in the Fee. Defined Cost includes only amounts calculated using rates and percentages stated in the Contract Data and other amounts at open market or competitively tendered prices with deductions for all discounts, rebates and taxes which can be recovered.
>
> **Schedule of Cost Components**
>
> An amount is included
>
> - only in one cost component and
> - only if it is incurred in order to Provide the Works.

Simply put, if the *Contractor* incurs any costs that are listed in the Schedule of Cost Components (SCC), they will be reimbursed by the *Client*. Costs that are not listed will not be eligible for reimbursement. In outline, that sounds simple. But, as with many things in life, this differentiation will provoke differing interpretations in the minds of different people and we find that many disputes concern the reimbursement of costs that *might* be reflected in the SCC.

> **Example 5.5 Schedule of Cost Components**
>
> - *Branded clothing.* The only clothing shown in the SCC that can be paid for is 'protective clothing'. If a *Contractor* wants its people to wear branded clothing, isn't that a matter for them? The *Contractor* may respond by claiming that all clothing on a construction site is protective, given the risk of skin cancer from exposure to the sun or similar conditions caused by exposure to, say, cementitious materials.
> - *Use of a road sweeper or vehicle wheel wash* by the *Contractor* to help maintain good relations with the local community. Where this provision is prescribed by the Scope then it is clearly necessary to Provide the Works, a condition precedent of any activity qualifying for reimbursement in the SCC. But what if the Scope is silent on the issue of keeping the adjacent roads clean? Is the use of a wheel wash necessary? The *Contractor* will say that it is, but the *Project Manager* may argue otherwise.
> - *Bonuses paid to people.* These are recognised in the SCC, but you must remember that any cost reimbursed must have been necessary to Provide the Works. On one occasion, we saw a significant bonus paid by the *Contractor* to a member of staff, ostensibly for her good work on the project in question. On being given documentary evidence of the payment of the bonus, the *Project Manager* saw, in the letter sent to the woman by her firm's chief executive, that the bonus reflected her 'many good years of service' to the company. In other words, the bonus was not wholly connected with the current project and therefore there was no reason why the current *Client* should fund all that bonus. On another occasion, a *Contractor* purchased a modest amount of takeaway food and drink from a well-known global fast-food restaurant to sustain a gang of concreters who had necessarily worked late one evening after a delay in their work. The *Contractor* claimed the reimbursement of this amount as a bonus; the *Project Manager* saw things differently and declined to certify that amount.

All three examples show that the SCC still leaves itself open to interpretation and parties can and do differ in their respective interpretations. Good drafting of any Z clause or ancillary documents helps minimise disagreements, but the primary recommendation we have for *Contractors* and *Project Managers* is to keep communicating with one another. Cost verification and interpretation of the contract require skilled people so thought should be given to who undertakes these tasks.

50. Defined Cost – dispute resolution

Our discussion of Defined Cost so far has touched largely on dispute avoidance, rather than dispute resolution. The efficient running of the contract by the *Project Manager* and the *Contractor* should involve creation of and access to documentary evidence. Where this is done properly, there should be a much smaller chance of a dispute.

But disputes do happen. Where the *Project Manager* and the *Contractor* do have to justify their actions, whether to the *Client* (for the *Project Manager*) or to an adjudicator, arbitrator or judge, they will be better placed to do so with comprehensive record documents. Where decisions to claim something, pay something or not pay something are concerned it will be useful to have contemporaneous file notes justifying those decisions. This makes explanation to a neutral person so much easier, particularly when, as is often the case, the explanation happens many months or years after the incident(s) in question.

Where a party to any contract has struggled to demonstrate something to its counterparty, it is almost certain that they will struggle more so explaining it to a third party, such as the *Adjudicator*.

The ECC's structure for dealing with Defined Cost, Disallowed Cost, Fees etc. provides a good framework from which the Parties can manage the reimbursement of costs. If that framework is followed and built on, disputes should be avoided and, if not, their resolution should be possible.

51. Main Option C – target contract with Activity Schedule

The remaining features of main Option C relate to its role as a target contract. Clause 11.2(32) defines the Prices as the 'lump sum prices […] on the Activity Schedule'. The Prices are the target that the Parties have agreed, most probably tendered competitively by the *Contractor*.

Although not defined, main Option C introduces the concept of the *Contractor's* share. The share is the difference between the Total of the Prices and the Price for Work Done to Date. Put simply, the difference between the target cost and the actual cost. The contract does not need to define the *Client's* share; it is simply the difference between the saving and the *Contractor's* share.

We discussed Activity Schedules earlier in this chapter under main Option A. There is no need to repeat that advice here; it is equally applicable to main Option C contracts.

But the Activity Schedule has a very different role here compared with that in main Option A; it sets the target. It is not used for the month-to-month payments process, so the granularity of its payments isn't as important as with main Option A.

52. Main Option D – target contract with Bill of Quantities

The remaining features of main Option D relate to its role as a target contract. The contract in Clause 11.2(33) defines the Prices as 'the lump sums and the amounts obtained by multiplying the rates by the quantities for the items in the Bill of Quantities'. The Prices are the target that the Parties have agreed, most probably tendered competitively by the *Contractor*. In this scenario, the target is affected by the actual quantity of work undertaken.

Although not defined, main Option D introduces the concept of the *Contractor's* share. The share is the difference between the Total of the Prices and the Price for Work Done to Date. Put simply, the difference between the target cost and the actual cost. The contract doesn't need to define the *Client's* share; it is simply the difference between the saving and the *Contractor's* share.

We discussed Bills of Quantities earlier in this chapter under main Option B. There is no need to repeat that advice here; it is equally applicable to main Option D contracts.

The Bill of Quantities has a very different role here compared with that in main Option B; it sets the target. It is still important that the Bill of Quantities is properly drawn up and properly administered. It is tempting, given the limited role of the Bill of Quantities in an Option D contract, to avoid measurement until Completion. But many building and engineering works are covered up immediately after construction (e.g. foundations, drains, ceiling voids); therefore, proper measurement must take place at the time to minimise disputes about computing the Total of the Prices. There is an additional defined term for Option D only; the Total of the Prices in Clause 11.2(35). This is the final measurement of the work and is used in Clause 54 when assessing the *Contractor's* share. The importance of maintaining an accurate measurement is paramount and can be seen from Clause 11.2(35) and its reference to 'the quantity of the work which the *Contractor* has completed'.

The resolution of any disputes that include arguments about what should have been measured will be complicated by the absence of agreed, accurate and contemporaneous records.

53. Main Option E – cost reimbursable contract

The remaining features of main Option E relate to its role as a cost reimbursement contract. Typically used in situations where the full extent of the Scope cannot be accurately predicted at the outset, it has a simpler payments process than the other main Options.

The Prices are defined in Clause 11.2(34) as 'the forecast of the total Defined Cost for the whole of the *works* plus the Fee'.

The term 'Prices' is used only in the aftermath of an implemented compensation event and allows the *Client* a forecast of the likely outturn cost for the *works*.

We can offer few other pieces of advice for this Option except for reiterating what we said earlier about Defined Cost. Where the Parties have agreed to share the risk of the final outturn cost of the project, they must work closely together to agree cost information, preferably at the time.

54. Main Option F – management contract

Option F is a management contract, where the *Contractor* does some activities for predetermined *prices* and where Subcontractors also undertake work. The payment provisions recognise this split. Some of the definitions and provisions used in Option F are identical or similar to those in Options C, D and E. But the overall position is different, so close attention is needed to the actual provisions, particularly by those who are familiar with the other Options.

The Price for Work Done to Date is familiar and is identical to earlier main Options.

> Clause 11.2(31)
>
> The Price for Work Done to Date is the total Defined Cost which the *Project Manager* forecasts will have been paid by the *Contractor* before the next assessment date plus the Fee.

However, Defined Cost has a different definition.

> Clause 11.2(25)
>
> Defined Cost is
>
> - the amount of payments due to Subcontractors for work which is subcontracted without taking account of amounts paid to or retained from the Subcontractor by the *Contractor* which would result in the *Client* paying or retaining the amount twice and
> - the *prices* for work done by the *Contractor*
>
> less Disallowed Cost.

Main Option F uses a new term, '*prices*', which isn't capitalised but is italicised. That avoids any conflict with the capitalised but unitalicised term 'Prices'. Still with us? Hopefully.

The *prices* are identified in Contract Data part two; that part of the document should be structured to reflect the activities required of the *Contractor* as set out in the Scope. Here the documents require careful drafting, with the *Contractor's* obligations written in clear and simple terms. If this is done, any debate about what the *Contractor* has done will be capable of being informed by reference to the contract's requirements.

The definition of Disallowed Cost differs from main Options C, D and E. Though there are lesser limbs of Disallowed Cost in Option F compared to main Options C, D and E, one additional provision is shown here.

> Clause 11.2(27) (excerpt)
>
> Disallowed Cost [...]
>
> - is a payment to a Subcontractor for
>   - work which the Contract Data states that the *Contractor* will do itself or
>   - the *Contractor's* management

So, the additional head of Disallowed Cost is if the *Contractor* pays a Subcontractor to do something that originally formed part of the workload included in the *prices*. Once again, this demonstrates the need to complete the Contract Data correctly. The entry in Contract Data part two for the 'Work which the *Contractor* will do' requires a list of activities with associated *prices*. This offers little scope for the *Contractor* to explain what its *prices* include. These should be explained in enough detail to avoid debate or dispute once those activities are concluded and payments are assessed. The *Contractor* is free subsequently to subcontract any of these activities but the costs of doing so will be disallowed.

55. X clauses

We explain the secondary Option clauses in Chapter 15. But it is worth mentioning here that many of them have an indirect impact on payments and some have a direct impact on payment assessments

X1	Price adjustment for inflation
X3	Multiple currencies
X6	Bonus for early Completion
X7	Delay damages
X12	Multiparty collaboration (specifically, the Key Performance Indicators)
X14	Advanced payment to the *Contractor*
X16	Retention
X17	Low performance damages
X20	Key Performance Indicators
X21	Whole life cost
X22	Early *Contractor* involvement.

It is important that the *Contractor* and the *Project Manager* anticipate the requirements for these clauses and include them in applications. Having agreed the inclusion of any secondary Options, both Parties must comply with them, even if, as for example, with delay damages, the impact is significant.

56. Schedules of cost components

There are two schedules of cost components in the ECC

- the Schedule of Cost Components (we now refer to this as <u>the</u> schedule)
- the Short Schedule of Cost Components (we now refer to this as the short schedule).

The short schedule is used to derive the Defined Cost when working with Options A or B. The Defined Cost in those Options is only used when assessing the financial impact of compensation events.

The schedule is used to derive Defined Cost when working with Options C, D, E or F. The Defined Cost in those Options is used to assess all payments.

We explained earlier that for an item in one of the schedules to be paid it must have been incurred in order to Provide the Works. Both schedules contain eight headings

1. People
2. Equipment
3. Plant and Materials
4. Subcontractors
5. Charges
6. Manufacture and fabrication
7. Design
8. Insurance.

Despite the NEC drafters' desire to be clear and simple, these sections are capable of different interpretations; therefore, the inclusion or exclusion of amounts from assessments should be clearly explained by the *Contractor* or the *Project Manager*. We gave some simple examples earlier of how even relatively small sums could cause difficulty for the Parties. The early planning of assessments should be undertaken by the *Project Manager* and the *Contractor*. What reports can the *Contractor's* accounts system provide? How will information be gathered, collated, allocated etc.? An effective assessment process can only work once the two protagonists work together, for example as they are required under Clause 20.4 in Options C, D, E and F.

The key issues for both Parties to remember is in Clause 52.1:

All the *Contractor's* costs which are not included in the Defined Cost are treated as included in the Fee.

The *Contractor* must understand this when pricing the *works* at the tender stage and subsequently when preparing payment applications.

Where NEC contracts reach the formal dispute resolution stage, the assessment of payments becomes very difficult, given the number of issues in the contract that are open to debate. The schedules of costs components are a significant part of these issues. The schedule contains three pages of text that, once the Parties have stopped collaborating, present several opportunities for disagreement. Avoiding these 'opportunities' is to be recommended. Parties' arguments in dispute resolution need to refer to the schedules' exact statements and explain why something is included or not as the case may be. References to what is normally included or excluded from Prices is irrelevant; sticking to the contract is always what an adjudicator, arbitrator or judge will be looking for.

57. Cost finalisation in main Options C, D, E and F

We have already explained cost finalisation (Clause 53.3) earlier in the context of the end of the project and the desirability for the Parties to conclude their work together. Similar provisions exist in the four cost reimbursable Options to finalise costs for a specified period of the project, most probably before

Completion. This might be for a financial period (such as the financial year) or a certain section of the *works*. The same principles apply, with the same benefits to the Parties.

Clause 50.9 exists in Options C, D, E and F and refers to a 'Defined Cost for a part of the *works*' being finalised. The 'part' is for the *Contractor* to decide when it submits the relevant data to the *Project Manager*. The *Project Manager* then has 13 weeks to accept that part of Defined Cost is correct or notify the *Contractor* that further work is needed on the assessment.

All NEC actions are time-limited in some way. But 13 weeks is a considerable period. If the *Project Manager* wants more information, this should be demanded quickly, and we suggest that this should happen much earlier than 13 weeks from submission. Where differences remain between the *Contractor* and *Project Manager*, they should still identify and record the costs on which they do agree, thus reducing the size of any dispute.

Unlike the provisions of Clause 53.3, Clause 50.9 only reaches a conclusion that part of the Defined Cost is 'correct'. This is a weaker conclusion than Clause 53.3, which provides 'conclusive evidence of the final amount due'. However, this process is to be commended in reducing the potential for disputes and contributes to reducing or avoiding a large quantum debate after Completion.

58. Z clauses

When working with standard forms of contract, amendments to adapt the form to the specific needs of the Parties and the project are common. NEC contracts refer to '*additional conditions of contract*' as Z clauses. Z clauses can remove or amend standard form clauses and add new ones. Therefore, their impact on payments must be analysed when drafting them and when using them. All NEC clauses have been written to dovetail with one another; Z clauses should be similarly written. Once again, this is a job for people who know what they're doing. We often see bad examples of drafting Z clauses where the drafter has omitted to allow for linkages between the Z clause and other provisions of the contract. We have the following advice for Z clauses, as with many other parts of the contract documents.

- Draft Z clauses carefully and with consideration for how the clause will interact with other parts of the contract.
- The *Contractor* and *Project Manager* should consider the impact of Z clauses when assessing payments due under the contract.
- Conflicts between the Z clause(s) and the standard provisions should be highlighted and resolved at the earliest opportunity.

NOTES
i   We understand that Clause Y(UK)2 also ensures compliance with similar legislation in Northern Ireland.
ii *Sutcliffe* v *Thackrah and Others* [1974] AC727.
iii *Costain & Ors* v *Bechtel & Anor* [2005] EWHC 1018 (TCC).
iv The 'black book' is the full version of the Engineering and Construction Contract which includes all six main options. A 'green book' contains just one main option; therefore, there are six green books, one for each of the main options A, B, C, D, E and F.
v   There are two sets of legislation, one for Scotland and one for England, Wales and Northern Ireland.
vi HMG (Her Majesty's Government) (1996) Housing Grants, Construction and Regeneration Act 1996. The Stationery Office, London, UK. Amended by HMG (2009) Local Democracy, Economic Development and Construction Act 2009. The Stationery Office, London, UK.

HMG (1997) The Construction Contracts (Northern Ireland) Order 1997. The Stationery Office, London, UK. Amended by HMG (2011) Construction Contracts (Amendment) Act (Northern Ireland) 2011. The Stationery Office, London, UK.
vii As defined in section 104(1) of the Housing Grants, Construction and Regeneration Act 1996, as amended.
viii ICE Publishing (2012) *CESMM4, Civil Engineering Standard Method of Measurement*. Thomas Telford, London, UK.
ix UK Department for Transport (2005) *Manual of Contract Documents for Highway Works*. The Stationery Office, London, UK, vol. 4.

Chapter 6
Compensation events

Compensation events are now the widely used C word in NEC contracts, replacing the previous C word that the original drafters of the ECC eschewed. The word 'claim' was deliberately avoided as it had generated connotations in the industry of a con, a try-on etc.

Compensation events are explained in the contract (the term isn't defined). Broadly speaking, they fall into three categories

- additional work instructed by the *Project Manager*
- the failure of the *Client*, *Project Manager*, *Supervisor* or Others to do something, either at all or on time
- the occurrence of a risk-based event, such as weather or physical conditions, which has exceeded the acceptable risk that the Parties had agreed the *Contractor* should bear.

As with all types of commercial contract, provisions that vary the amount paid and potentially lengthen the duration will attract attention and will lead to disputes, even where the provisions are followed. In the ECC, the compensation event provisions are linked to many other parts of the contract; payments, programmes, early warnings, communications etc. So, understanding the detailed requirements of those other parts of the contract is just as important as knowing what to do to operate the compensation events processes properly.

The compensation event provisions provide a structured approach to the management of change; dealing with the financial and time aspects simultaneously. The four-step process is intended to provide a clear route to the implementation of change in the contract and should, if necessary, be capable of being concluded by one Party without the other's participation. Clearly, that scenario isn't desirable but it needs to be provided for. The most desirable scenario is where the Parties agree on the consequence of change. But agreement isn't necessary to implement a compensation event.

The finality of a compensation event is stated in Clause 66.3. If revision is needed after implementation, the *Project Manager* is unlikely to have a further role, as the *Senior Representatives* and then the *Adjudicator* are the next people to involve when using Clause W1 or W2.[i] Alternatively, the Parties can make an agreement subject to Clause 12.3 to vary the assessment, but this is rare.

The four stages of the process are as follows.

- Notification, where the occurrence of a compensation event must be recognised and notified to the other Party.
- Quotation, where the *Contractor* will provide its own assessment of the contractual impact of the event, in accordance with the contract's assessment rules.

- Assessment, where the *Project Manager* will establish whether the *Contractor* has provided a correct quotation and, if the *Contractor* has not done so, will make an assessment, which will then be implemented.
- Implementation, where the *Contractor's* quotation (if accepted by the *Project Manager*) or the *Project Manager's* assessment will lead to the changing of the Prices, the Completion Date and the Key Dates, where appropriate.

59. Where are compensation events described in the contract?

Compensation events are found in five places

- core clauses
- main Option clauses
- secondary Option clauses
- Contract Data (if used, via Clause 60.1(21))
- Z clauses.

With the addition in NEC4 ECC of Clause 60.1(21), additional compensation events created through Z clauses are likely to be rare. Z clauses are more likely to remove or amend compensation events from the standard form, in our experience.

The primary source of compensation events is Clause 60.1(1), where 20 types of compensation event are listed, together with a cross-reference to any additional compensation events specified in the Contract Data.

Further compensation events are also listed in

- Clauses 60.4, 60.5, 60.6 (main Options B and D) (these refer to issues with the Bill of Quantities)
- secondary Option Clauses X2.1 (changes in the law), X12.3(6) and (7) (multiparty collaboration), X14.2 (advanced payment), X15.2 (the *Contractor's* design), Y2.5 (suspension).

The ability to add compensation events is provided in Clause 60.1(21) and also with Z clauses to allow the Parties (albeit, it will be the *Client's* decision most of the time) to amend the risk profile of a contract. Why might the *Client* do this? We have seen this facility used where a project faces unpredictable risks, for example the treatment of contaminated materials, such as asbestos, or protester or security problems. These risks might lead to additional time and cost that is impossible to estimate in advance. To ask the *Contractor* to face such risks is likely to lead to significant additional premiums being added to the tender figures, potentially skewing the tender appraisal. The *Client* will end up paying for these risks even if they don't materialise. So, by adding an additional compensation event(s), the *Client* will remove such risk from the *Contractor* and will only face additional cost if the risk materialises.

The addition of further compensation events provides additional scope for disputes. As with all bespoke additions or amendments to the contract, the wording should be carefully written by someone who is appropriately skilled in drafting and who also understands the standard version of the contract. Additional compensation events need to dovetail with the rest of the contract. Legal interpretation tends to favour the amended provisions over the standard wording where a conflict between the two is found. Parties contracting based on an amended term need to have certainty as to what their respective obligations and entitlements are.

We now describe the compensation events in turn and the issues that the Parties might face in dealing with them.

> **Clause 60.1(1)**
>
> The *Project Manager* gives an instruction changing the Scope except
>
> - a change made in order to accept a Defect or
> - a change to the Scope provided by the *Contractor* for its design which is made
>   - at the *Contractor's* request or
>   - in order to comply with the Scope provided by the *Client*.

In our experience, this is probably the single most common compensation event and consequently one of the most common to feature in dispute proceedings. *Project Manager's* instructions should, of course, comply with Clause 13 but in many projects changes to the Scope come in a variety of guises, some oral, some written, some less obvious than others. Revisions of drawings are a common source of disagreement; is the amendment a change to the Scope, or just design development? The issue of design development has caused difficulties with projects for many years but it is important to remember that the phrase isn't mentioned in the ECC, nor is it inferred in any way. It is important for both the *Contractor* and the *Project Manager* to remain alert to changes in the Scope, large and small, and to alert one another through the notification provisions in Clause 60.1. Where the *Project Manager* is aware of a change and hopes that the *Contractor* won't spot it, those hopes are normally dashed.

For changes that do lead to a compensation event, there needs to be an instruction from the *Project Manager* that can be clearly identified. Not all such changes are accompanied by an explicit instruction. Should there be a letter? A transmittal note? An email? Is an automated notification from a communication system stating that a new document has been uploaded enough? These are things for the Parties to work out, well before the subtleties become important.

Two exceptions are given to this compensation event. The first is where the change has been made in order to accept a Defect. There is a direct link here with Clause 45.2, although that clause refers to the *Project Manager* changing the Scope, not to the giving of an instruction. Careful documentation of the correspondence required by Clauses 45.1 and 45.2 will assist in avoiding any misunderstanding that subsequently turns into a dispute.

The second exception is where the *Contractor's* Scope for its design is changed at the *Contractor's* request or to comply with the *Client's* Scope. These two are relatively clear; it is this point that ensures the primacy of the *Client's* Scope over that of the *Contractor*. The wording is careful though. If the *Client's* Scope is changed to match the *Contractor's*, a compensation event will ensue. Again, careful documenting of the changes will be needed.

> **60.1(2)**
>
> The *Client* does not allow access to and use of each part of the Site by the later of its *access date* and the date for access shown on the Accepted Programme.

Access to the Site is clearly necessary for most construction projects even if, say, design and fabrication activities are taking place elsewhere. Note the distinction between the Site and the Working Areas; this compensation event only refers to the Site.

This compensation event reinforces the need to get two things right.

- The Contract Data needs to contain the *access date(s)* and the *Client* needs to be able to honour them.
- The Accepted Programme must be administered properly so that the existence of a compensation event can be demonstrated. *Access date(s)* and the date(s) for access need to be shown on the programme, as can be seen from Clause 31.2. The *Contractor* should show this information and the *Project Manager* should ensure that the dates are compliant before accepting the programme. Where Sites are subdivided into different sections, each with its own *access date*, further complications can arise. If a number of dates are to be used, the parts of the Site to which they refer must be spelt out clearly.

> **Example 6.1**
>
> A regional municipal waste collection and treatment project required *works* on 52 Sites. Six were major treatment plants, the remainder were local household recycling centres. Each Site had its own access issues and planning challenges. Some Sites were leased, while some relied on third-party land for access. The potential for access-related delays was therefore significant.
>
> Delays were encountered. But the prior hard work of the Parties ensured that the identification and assessment of these issues was not over-complicated. The contract required that the *Contractor* and *Project Manager* utilised the same software for programming. The Contract Data contained clear descriptions of each location, together with the appropriate *access dates*.
>
> The two Parties monitored progress against each date and when access would be needed. Using different levels of granularity in the programmes allowed each individual delay to be assessed at a local level and for the project overall. No disagreements were encountered because of the way the contract documents had been drafted and of how the *works* were managed in a collaborative manner.

> **Clause 60.1(3)**
>
> The *Client* does not provide something which it is to provide by the date shown on the Accepted Programme.

For this compensation event to exist, the *Client* must have had an obligation to provide something that was contractually binding. There is a link here to Clause 25.2, which states that such obligations will be specified in the Scope. If time is critical to the provision of such 'services and other things', the dates should also be specified in the Scope; then this can be reflected by the *Contractor* in the programme.[ii]

If the *Client's* provision is late or non-existent, this needs to be measured against the Accepted Programme. When deciding whether to accept the programme, the *Project Manager* needs to check that the *Client* can provide what is needed on the dates claimed by the *Contractor*. Late provision under this compensation event is in respect of the Accepted Programme, not the contract.

Clause 60.1(4)

The *Project Manager* gives an instruction to stop or not to start any work or to change a Key Date.

The *Project Manager* wields considerable authority. This compensation event links to Clause 34.1, which provides the *Project Manager* with the ability to instruct interruptions to the *works*. The important advice in this type of situation is for the *Project Manager* to comply with the contract's communications clauses and to state clearly what activity is subject to the instruction. Good records must be kept by both the *Contractor* and the *Project Manager*. What has happened during the stoppage? Which resources were affected? What knock-on impact might the stoppage have? All these things must be documented at the time. While this compensation event doesn't need to be notified by the *Contractor* to avoid the time bar in Clause 61.3, it is advisable that the *Contractor* does notify the *Project Manager* as soon as becoming aware, if a notification is not received from the *Project Manager*.

Clause 60.1(5)

The *Client* or Others

- do not work within the times shown on the Accepted Programme,
- do not work within the conditions stated in the Scope or
- carry out work on the Site that is not stated in the Scope.

This compensation event relates to Clauses 31.2, 25.1 and 25.2. The *Contractor* is entitled to rely on the information in the contract and in the Accepted Programme. Note again the reference to the Accepted Programme in one list item and references to the Scope in the remaining two. This clause must be interpreted carefully; the reference to the Accepted Programme is another warning for the *Project Manager* when checking a programme before accepting it. Can the *Client* or Others, or both, comply with the times stated in the programme by the *Contractor*? Once a programme is accepted by the *Project Manager*, the *Client* is at risk for its own failure to perform to those times and for the failure of Others too, so the *Project Manager's* vigilance is vital.

The probable outcome of this compensation event is delay or disruption, or both. Disruption claims, as distinct from delay claims, are difficult to demonstrate and evaluate under any form of contract and the ECC is no different. Once again, the Parties can make things easier for themselves by keeping accurate records of the actual effects of events. Forecasting the effects of compensation events, such as this one (as required by Clause 63.1) is not likely to be straightforward, so records will be doubly important.

> **Clause 60.1(6)**
>
> The *Project Manager* or the *Supervisor* does not reply to a communication from the *Contractor* within the period required by the contract.

So why have the NEC's drafters included this clause? Many of the important clauses contain time limits for the *Project Manager* to do certain things and the Scope is likely to contain more for the *Supervisor* too. The failure of anyone in a project to reply to communications can be frustrating for the innocent party, but it isn't always easy to compute any financial or time impact of the failure.

Where the *conditions of contract* and the Scope contain time restrictions, there is often a remedy for the failure to respond on time. For example, the compensation event provisions and programme submissions allow for the *Contractor* to submit a reminder to the *Project Manager* and for the relevant submission to be treated as accepted after two weeks. This compensation event in Clause 60.1(6) then provides for the effect of the delay that is incurred in that situation.

There is a similarity between this compensation event and that in Clause 60.1(9), the latter relating to the *Project Manager's* withholding of an acceptance. This compensation event is broader and covers a range of communications. Clause 60.1(9) refers to the withholding of acceptance, which covers a smaller number of breaches. It could be argued that Clause 60.1(9) only applies when the *Project Manager* states that it does not accept a submission, as distinct from remaining silent. But more of that later.

There is also a similarity between this compensation event and that in Clause 60.1(11), the latter referring to the *Supervisor* causing unnecessary delay in tests and inspections. Clearly, where the *Supervisor* does not reply to communications about tests and inspections then unnecessary delay is likely to ensue.

Co-operation between the *Contractor*, *Supervisor* and *Project Manager* should minimise the occurrence of this compensation event. But, if such collaboration is rare, then there is a risk that some communications will be answered tardily and that compensation events will follow.

> **Clause 60.1(7)**
>
> The *Project Manager* gives an instruction for dealing with an object of value or of historical or other interest found within the Site.

The discovery of valuable objects or historical items typically involves other legal considerations in most legal jurisdictions; the Parties to an ECC contract will not have the freedom to decide what to do when, say, human remains are found. Usually the Police are the first port of call when this happens, and they will hand over to archaeologists once criminal conduct has been ruled out.

This compensation event is not triggered by the discovery of such items but instead by the *Project Manager* giving an instruction. Therefore, there is an assumption that the *Project Manager* will act quickly when such things are discovered.

If the *Project Manager* did not provide the necessary instruction and the *Contractor* were to continue works that could damage the items discovered, this would possibly be illegal; therefore, a notice under Clause 17.2 might prompt the *Project Manager* into action.

As with other situations involving delay and disruption, good records should be retained by all concerned; this will make subsequent administration and (where necessary) dispute resolution easier.

> Clause 60.1(8)
>
> The *Project Manager* or the *Supervisor* changes a decision which either has previously communicated to the *Contractor*.

This clause has a debatable impact; it isn't immediately clear what it refers to as 'decision' is not a defined term and therefore should be considered to have its ordinary meaning. In most projects, the *Project Manager* and *Supervisor* will make many thousands of decisions; almost every acceptance, certificate etc. involves a decision. Therefore, it might be argued that each change of mind could be a compensation event under this clause. We suggest that such a broad interpretation is unlikely to be effective here and that the clause only operates where the contract contains an express obligation for the *Project Manager* or *Supervisor* to decide something or issue a decision. There are no such obligations for the *Supervisor* in the *conditions of contract*, but the Scope might contain such references if the drafters have used the words 'decide' or 'decision'.

The *conditions of contract* have six references to the *Project Manager* deciding something

Clause 25.3   The achievement of a Condition for a Key Date
Clause 30.2   The date of Completion
Clause 50.1   First assessment date
Clause 61.5   *Contractor's* failure to give an early warning
Clause 64.1   *Contractor's* failure to assess a compensation event correctly
Clause 63.17 (main Option F only)   Changes to Prices and time.

There is also a reference to decisions in early warning meetings, but they are not made unilaterally by the *Project Manager*; rather, they are made in collaboration. So, we consider that a decision made at an early warning meeting is unlikely to lead to a compensation event if it is subsequently changed.

The revision of decisions by either the *Project Manager* or the *Supervisor* should be carefully documented. The wording of the new decision and any analysis of its contrast from the original one should be carefully drafted with the active consideration of how it might be interpreted in a dispute situation.

> **Clause 60.1(9)**
>
> The *Project Manager* withholds an acceptance (other than acceptance of a quotation for acceleration or for not correcting a Defect) for a reason not stated in the contract.

We referred to this compensation event earlier when comparing its use with that of Clause 60.1(6), the failure of the *Project Manager* or the *Supervisor* to reply to a communication. This clause only refers to the *Project Manager's* withholding of an acceptance so, if the Scope refers to the *Supervisor* having to accept something, this compensation event will not apply.

The *Project Manager* accepts or does not accept many things in the ECC and we have listed them all below to alert readers to the myriad of different situations that could be caught by this compensation event.

It is worth referring too to Clauses 13.4, 13.8 and 14.1 to demonstrate the wide-ranging ability of the *Project Manager* to decline acceptance of something. These clauses regulate the *Project Manager's* conduct in these matters and explain that where acceptance is withheld for a reason not stated in the contract, a compensation event will ensue.

The items for the *Project Manager* to accept (or not as the case may be) are

Clause 16.2	*Contractor's* proposals
Clause 16.3	Adding to the Working Areas
Clause 21.2	Design (of permanent *works*)
Clause 23.1	Equipment design
Clause 24.1	Replacement of *key person*
Clause 26.2	Subcontractor
Clause 26.2	Subcontract documents
Clause 31.1	Programmes (and Clauses 31.3 and 32.2)
Clause 40.2	Quality plan and quality policy statement
Clause 50.9	Defined Cost finalisation (main Options C, D, E and F)
Clause 55.3	Revised Activity Schedule (main Option A)
Clause 62.3	Compensation event quotation
Clause 65.2	Proposed instructions
Clause 84.1	Insurance certificates
Clause X4.2	Alternative guarantor for ultimate holding company guarantee
Clause X10.1	Information Execution Plan
Clause X13.1	Bank or insurer to provide performance bond
Clause X14.2	Bank or insurer to provide advance payment bond
Clause X16.3	Bank or insurer to provide retention bond
Clause X21.3	Whole life cost quotation
Clause X22.2	Defined Cost forecasts
Clause X22.3	Design proposals for Stage Two

Clause Y1.4   Banking arrangements for Project Bank Account
Clause Y1.6   Adding a Supplier to the Named Suppliers.

This list provides a useful explanation of just how many issues the *Contractor* must involve the *Project Manager* with. The blanket rule in Clause 13.4 requires the *Project Manager* to state 'the reasons in sufficient detail' when not accepting something. Those reasons can therefore be considered by the *Contractor* and an assessment can be made as to whether the *Project Manager* has effectively generated a compensation event. Of course, such interpretations can be subjective and a lack of description by the *Project Manager* can make things difficult. For example, the *Project Manager* may not accept a design submission, stating that it doesn't comply with the applicable law. But if the legal point is not explained in enough detail, as required in Clause 13.4, the *Contractor* will not be able to assess whether the *Project Manager* is correct or not and will most likely proceed on the basis that the *Project Manager* is not correct. This type of communication leads to a lot of festering problems, which, with time, will lead to disputes.

Avoiding disputes here requires the following.

- The *Contractor* should provide enough detail of what is being submitted for acceptance.
- Where the *Project Manager* does not accept the submission, the accompanying explanation should be clear and sufficiently detailed to explain the methodology.
- The *Project Manager* and *Contractor* should meet to discuss items that have not been accepted and allow Others to attend, particularly if an early warning has been given and an early warning meeting is being held.

The two exceptions in this compensation event refer to the actions in Clauses 36.1 and 45.2. The acceptance of an acceleration quotation and of defective work is exempt from time restrictions, as these are optional actions, which the *Project Manager* and *Contractor* are free to use or avoid as they wish.

Clause 60.1(10)

The *Supervisor* instructs the *Contractor* to search for a Defect and no Defect is found unless the search is needed only because the *Contractor* gave insufficient notice of doing work obstructing a required test or inspection.

This compensation event reflects the *Supervisor's* authority in Clause 43.1 to instruct the *Contractor* to search for a Defect. A lot of construction work gets covered up by succeeding work, for example drainage pipes in trenches, brick ties in cavities, building services in ceiling voids and so on. Therefore, where the *Supervisor* has concerns about defective work, an instruction may be needed to uncover work. The compensation event does not occur in the following situations.

- Where a Defect is found. It does not have to be the Defect specified by the *Supervisor* (emphasis added).
- Where the *Contractor* had given insufficient notice of doing work; in other words, where the *Supervisor* was not afforded the opportunity to inspect work prior to it being covered up.
  For this to work effectively, the Scope must contain detailed procedures for the co-ordination of the *Contractor's* and *Supervisor's* duties for testing and inspection. Items such as hold points and notice periods must be carefully identified.

Instructing the *Contractor* to search for a Defect can involve dismantling or even destroying completed work so the *Supervisor* must ensure that this is a last resort and that no other route exists to satisfy a concern. For example, quality records, photographs or CCTV images could all be used to analyse the previous installation processes.

> **Clause 60.1(11)**
>
> A test or inspection done by the *Supervisor* causes unnecessary delay.

This compensation event is like the previous one, in that we can minimise its occurrence by drafting the Scope carefully. All tests and inspections will cause delay, but presumably the tests and inspections are necessary, so the delay is too. This compensation event matches the wording in Clause 41.5. By drafting the Scope properly, the Parties can ensure that the *Contractor* and the *Supervisor* know of each other's requirements for testing and inspection. They are required to communicate with one another under Clause 41.3.

To avoid disputes under this clause, the *Supervisor* and the *Contractor* should ensure the following.

- The *Supervisor* only does those tests recognised in the *conditions of contract* and the Scope and does not demand testing or performance of a higher quality merely as a matter of personal preference.
- The proper notice periods are given by the person(s) undertaking the test to the other person(s).
- Close co-ordination is undertaken and those involved remain flexible to each other's needs.

> **Clause 60.1(12)**
>
> The *Contractor* encounters physical conditions which
> - are within the Site,
> - are not weather conditions and
> - an experienced contractor would have judged at the Contract Date to have such a small chance of occurring that it would have been unreasonable to have allowed for them.
>
> Only the difference between the physical conditions encountered and those for which it would have been reasonable to have allowed is taken into account in assessing a compensation event.

This compensation event is a common one and the situation in question affects many construction contracts. Typically, the wording refers to ground conditions being more expensive and slower to work with than some level of expectation of the Parties at the outset. Physical conditions, particularly those below ground, are difficult to predict even where there has been some Site investigation prior to the Contract Date. Actual conditions may and often do differ from those predicted conditions, leading to additional costs and delay. Material predicted to be soil may turn out to be rock when excavated, water may be experienced at a higher level than predicted and so on.

For the *Client*, these claims are difficult to explain to a lay audience. More expenditure on foundations creates no additional value in whatever sits on top of those foundations, so claims of this nature often lead to a puzzlement in those sanctioning the payments. Unsurprisingly, these types of experience can lead to Z clause amendments in future contracts, seeking to remove the provision entirely or amend it. Where this clause is removed, it doesn't transfer the risk of physical conditions entirely to the *Contractor* as, particularly where the design is provided by the *Client*, instructions by the *Project Manager* may be required to overcome the conditions found and a compensation event will then occur because of those instructions.

This clause also links to Clauses 60.2 and 60.3 and to the Site Information.

Looking at the provisions overall, the Parties need to make the following considerations.

- The conditions experienced must be within the Site. If they are found elsewhere in the Working Areas then, whatever the impact, there won't be a compensation event.
- They must not be weather conditions. Rain and snow are physical conditions but are covered in Clause 60.1(13). Occasionally, there are disputes about flooding: is that a physical condition, or a consequence of weather conditions?
- The test as to whether an <u>experienced</u> contractor, as distinct from <u>the</u> *Contractor*, would have judged this to be a sufficiently large risk to have allowed for it is a subjective one. As with every other test in the ECC, what <u>the</u> *Contractor* allowed for is immaterial. The drafters have tried to create an objective test here but its interpretation will inevitably lead to different views. Once physical conditions are encountered that lead to a compensation event, the *Project Manager* and *Contractor* will pore over the information available at the tender stage to suggest what an experienced contractor would have allowed for. Differing opinions are common in this analysis.
- Where it is demonstrated that the physical conditions were different from those that an experienced *Contractor* would have allowed for, it is only the effect over and above that level that is compensated for.
- Clause 60.2 sets out several assumptions as to the level of knowledge that the *Contractor* is assumed to have had when deciding what to allow for in the contract. It is still a feature of many disputes that the *Contractor* has not visited the Site before the Contract Date. We have seen a contractor surprised to discover water ingress while excavating in gravel next to a river and another contractor struggling with the removal of rock just below the surface when working on a hillside punctuated with visible rocky outcrops. Physical conditions are a difficult enough subject, without the *Contractor* failing to avail itself of the information available.
- Clause 60.3 provides some direction in the event of an ambiguity or inconsistency, but this is only in relation to the contents of the Site Information and not any difference between the Site Information and other documents. This latter situation would be dealt with under the remaining provisions of the contract.

The additional works caused by these events tend to be demanding of resources, particularly Equipment and people. Detailed records are needed of the impact of physical conditions; these must be maintained by the *Project Manager* and the *Contractor*. Ideally the records will be agreed while they are contemporaneously prepared.

Avoiding disputes in this regard can be difficult, owing to the often-large sums of money involved and the subjective nature of the tests needed to identify liability. But the resolution of disputes will be simplified if there is some agreement about the costs incurred in dealing with those conditions.

> **Clause 60.1(13)**
>
> A *weather measurement* is recorded
> - within a calendar month,
> - before the Completion Date for the whole of the *works* and
> - at the place stated in the Contract Data
>
> the value of which, by comparison with the *weather data*, is shown to occur on average less frequently than once in ten years.
>
> Only the difference between the *weather measurement* and the weather which the *weather data* show to occur on average less frequently than once in ten years is considered in assessing a compensation event.

The issue of weather has caused difficulty in construction projects since records began, or certainly records that survive to this day. Even with modern-day construction becoming increasingly reliant on offsite manufacturing, there will always be work to do outside. Wherever work takes place in the world, contractors must plan for routine weather conditions, but it is the extremes that cause difficulties in relationships.

Other forms of contract provide relief for 'exceptionally adverse weather' or other such words. Such subjective phraseology has created difficulties over the years and can contribute to disputes. One person's exceptionally adverse rain is another's predictable rainfall. The ECC has a mechanism designed to avoid such subjective arguments, one that should produce a binary result.

As with all the compensation events, the wording must be read carefully. The mechanism provides compensation to the *Contractor* for weather conditions that are in excess of the stated threshold. But the comparison is month by month. In other words, the *Contractor* needs to experience weather worse than the ten-yearly average in, say, November this year.[iii] The actual weather measurements for November this year will be compared with the ten-yearly average for all Novembers on record. If the *Contractor* suffers 31 days of bad weather from, say, 15 November to 14 December, this might not qualify as a compensation event if milder conditions were experienced in early November or late December.

The poor weather must be experienced before the Completion Date. So, if the *Contractor* is running late, it won't receive any compensation for the effects of the weather, nor will it gain relief from delay damages.

And, of course, the weather must be experienced at the place stated in the Contract Data. Often this is the Site, if no nearby and relevant monitoring facilities exist already.

Those negotiating or setting contract conditions should consider how weather will affect the *works*. Wind will affect cranage and will bring lifting operations to a standstill once the wind speed exceeds the maximum allowed under health and safety regulations. But wind conditions aren't listed in the Contract Data of the standard ECC form, so the Parties will need to create their own entries. Wind speed is usually measured in the use or not of cranes, so consider if an automated measurement and recording system can be established to ensure that both Parties are working from the same source of data. Wind is usually a very localised problem, owing to the local topography, so generic wind speed results are unlikely to be relevant. Where we do see wind-related compensation events, we typically see Z clauses deployed to create bespoke risk-sharing provisions. As with all bespoke amendments, these must be written to dovetail with the original wording.

The Contract Data form does include entries for the effects of rainfall and temperature extremes. Extreme high or low temperatures affect the ability to undertake wet trades, such as concreting, plastering and brickwork. Rain affects the ability to install materials that require compaction or high-quality surface finishes. Of course, these risks can and should be mitigated by the provision of shade, shelter, heating and so on. And all these weather extremes affect productivity because they impact on the ability of human beings to work manually. Perhaps one of the best arguments in favour of factory-based manufacture?

Working with weather data requires

- thought being applied when the contract is drafted
- consideration of the effects of the weather risks by tenderers
- monitoring of weather data by both Parties and by the *Project Manager* and the *Supervisor*
- careful interpretation of the data by the *Contractor* and the *Project Manager*, although it is usually the former that is keener to undertake this role, particularly as this compensation event will need to be notified in most situations by the *Contractor*
- co-operation between the *Contractor* and the *Project Manager* for assessing the impact of the compensation event.

The assessment of this compensation event requires the Parties to judge the marginal increase in cost and time experienced by the *Contractor* over and above the ten-yearly measure that the contract requires the *Contractor* to provide for. So, if the *Contractor* has experienced five days of excessive rainfall during November and the historic ten-yearly measure states that only two days of rainfall of that level would have been experienced, compensation is due for the effect of the three additional days.

> Clause 60.1(14)
>
> An event which is a *Client's* liability stated in these *conditions of contract*.

This clause relates to the *Client's* liabilities listed in Clause 80.1. That list does not require explanation here, but this clause ensures that the compensation provisions and assessment rules will apply to the occurrence of any of the *Client's* liabilities. The liabilities can be extended by the final list item in Clause 80.1, which cross-refers to the Contract Data.

We cannot provide a description of the likely circumstances of each of the *Client's* liabilities listed in Clause 80.1, but our experience suggests that the most common one is a fault in the *Client's* design. This is a difficult subject, given the creative nature of design. Demonstrating that design is faulty can be a difficult task, and forecasting its effects (as per Clauses 63.1 and 63.3) even more so.

> **Clause 60.1(15)**
>
> The *Project Manager* certifies take over of a part of the *works* before both Completion and the Completion Date.

This compensation event rests on the *Project Manager* issuing a certificate, as opposed to the action of take over itself. Most people do not read the last part of this clause either, in that this has to happen <u>before</u> both Completion <u>and</u> the Completion Date. The link here is to Clause 35.3, which requires the *Project Manager* to certify the date on which the *Client* takes over any part of the *works*. The *Project Manager* must certify within one week of take over. So, we have the possibility that take over may occur and that the *Project Manager* may be tardy in the certification or perhaps not do it at all. In that situation, the *Project Manager* would be in breach of Clause 10.1 by not acting as stated and there isn't an express remedy in the contract for the *Project Manager's* failure in this regard, aside from following the dispute resolution procedures.

The need for this compensation event is to reflect the disruption and additional costs that might be incurred by the *Contractor* if the *Client* 'moves in' and starts using the *works* while the *Contractor* is still trying to construct them, having not been allowed the agreed time period to reach Completion. Once the Completion Date has passed, there will not be a compensation event as the *Contractor* will be in breach and the *Client's* actions will be mitigating its own losses as a result of the delayed Completion.

If the contract's programming obligations have been complied with, any early take over by the *Client* will have been expected and planned for; this would hopefully reduce the time and cost impact of such actions.

> **Clause 60.1(16)**
>
> The *Client* does not provide materials, facilities and samples for tests and inspection as stated in the Scope.

This compensation event relates to the *Client's* (reciprocal) obligations in Clause 41.2 to provide these things as stated in the Scope. This is like the compensation event in Clause 60.1(3); that differs from this one in its reference to the Accepted Programme. Here, the reference is to the Scope. Some sites may benefit from laboratory services provided by the *Client* or the *Supervisor*; typically, we see these stated in the Scope. Where they are subsequently not provided, there will be a consequential impact on costs and possibly timing too. Where samples are not provided, these will cost money if the *Contractor* must acquire them. Any instruction to provide such things may trigger a compensation event under Clause 60.1(1).

As with most compensation events, good records will assist with the assessment of this type of incident, hopefully avoiding any dispute.

> **Clause 60.1(17)**
>
> The *Project Manager* notifies the *Contractor* of a correction to an assumption which the *Project Manager* stated about a compensation event.

This compensation event links to the *Project Manager's* ability in Clause 61.6 to state assumptions before instructing the *Contractor* to submit a quotation for a compensation event. Where there is some doubt, the *Project Manager* can state an assumption. For example, will additional work be undertaken overnight on a weekend or during normal working hours midweek? The answer to this question will have a bearing on the costs of people and Equipment. The *Project Manager*, having stated assumptions, takes the risk of uncertain costs away from the *Contractor*. If, once the event has happened, the assumptions turn out to be incorrect, then a further compensation event will occur to correct the effect on the Prices and dates of the error.

The original assumptions and the later correction should be clearly described by the *Project Manager* to aid the assessment of the correcting compensation event under this clause. Depending on the cause of the original compensation event, the correcting compensation event may lead to a reduction in the Prices.

Clause 60.1(18)

A breach of contract by the *Client* which is not one of the other compensation events in the contract.

This clause leads to the compensation event provisions dealing with the consequences of an otherwise-unspecified breach by the *Client*. It needs to be read in conjunction with Clause 63.6, which sets out to create the compensation event provisions as a sole remedy for both Parties. Whether the drafting successfully achieves this end is probably best answered by those with legal qualifications, so we will comment no further.

We think that the aim of this clause is to provide a mechanism for assessing the effect of a *Client* breach that isn't otherwise included in the drafting. The intent is probably to avoid the quantum of such a breach being entirely at large. Given our concerns explained earlier, we think that the Parties should take advice before investing too much effort in this compensation event, should a breach be alleged.

Clause 60.1(19)

An event which

- stops the *Contractor* completing the whole of the *works* or
- stops the *Contractor* completing the whole of the *works* by the date for planned Completion shown on the Accepted Programme,

and which

- neither Party could prevent,
- an experienced contractor would have judged at the Contract Date to have such a small chance of occurring that it would have been unreasonable to have allowed for it and
- is not one of the other compensation events stated in the contract.

This compensation event is sometimes referred to as the 'force majeure clause'. The term 'force majeure' doesn't have a defined meaning so that perhaps doesn't help much in understanding this clause.

The first list refers to the event stopping the Completion of the *works*. But on closer reading, the second item in this list is effectively referring to something that slows down the achievement of Completion, by reference to the Accepted Programme.

For this compensation event to be triggered, the second list must be satisfied. Events of this nature tend to be rare (second list item), an example being extreme natural conditions, such as flooding or, as happened in the UK in 2001, an agricultural disease (foot-and-mouth disease) epidemic that caused large areas of land to be inaccessible. Finally, the event must not be one of the other compensation events in the contract.

The unpredictable nature of events such as these means that, more than ever, good records must be kept. The industry in the UK struggled to deal with the commercial aftermath of the foot-and-mouth outbreak in 2001, owing to the 'act now, record later' nature of events. That is possibly understandable given the unique circumstances, but some thought should be given to these issues at the time.

> **Clause 60.1(20)**
>
> The *Project Manager* notifies the *Contractor* that a quotation for a proposed instruction is not accepted.

This compensation event links to Clause 65, which provides for the *Project Manager* to instruct the *Contractor* to submit quotations for proposed instructions but where the *Contractor* does not immediately put an instruction into effect, thereby avoiding the need to comply with Clause 27.3. This provision allows the *Project Manager* (presumably on behalf of the *Client*) to explore the effect of a proposed instruction without committing to it. If the quotation is attractive, then the *Project Manager* can instruct the works (Clause 65.2). If not, the *Project Manager* must notify the *Contractor* that the quotation is not accepted and notify a compensation event for the effect of preparing the (abortive) quotation.

The effect here is to compensate the *Contractor* for the time and cost of preparing a quotation; sometimes this may be considerable if design and planning are required. The clause will also protect the *Contractor* against being used as a free estimating service and being distracted away from the 'day job'.

Demonstrating the cost may prove challenging in some situations, particularly if the people preparing the quotation are already paid for under Defined Cost applications. Good records, as ever, are important. We describe the operation of Clause 65 later in this chapter.

> **Clause 60.1(21)**
>
> Additional compensation events stated in Contract Data part one.

We explained earlier in this chapter about the need for good drafting when amending or adding new compensation events. This clause allows additional compensation events to be stated in Contract Data part one. The wording is specific to part one, thus avoiding the potential for an erroneous entry in part two by a tenderer that could lead to an unintended change in risk profile.

Additional compensation events may be desirable; we provided some examples earlier.

60. Additional compensation events in main Options B and D

There are further compensation events that only apply to contracts using main Options B and D, where a Bill of Quantities has been used. These recognise that the Bill of Quantities is usually prepared by the *Client* and that the risk in its accuracy therefore rests with the *Client*.

The correct preparation and use of a Bill of Quantities requires considerable professional skill; it is our experience that these skills are less common in the industry than they once were, owing to the reduced use of bills in construction contracts. Ironically this is probably something caused partly by the success of NEC contracts, where only a third of the main Options involve a Bill of Quantities.

Despite the welcome advance of technology and the part-automation of taking off quantities from designs, we still need people with an understanding of how bills should be used and the role of a *method of measurement*. The following three additional compensation events require administration by people with the relevant skills. Bills of quantities can appear to be deceptively simple, yet the consequences of getting them wrong can be considerable.

> **Clause 60.4**
>
> A difference between the final total quantity of work done and the quantity stated for an item in the Bill of Quantities is a compensation event if
>
> - the difference does not result from a change to the Scope,
> - the difference causes the Defined Cost per unit of quantity to change and
> - the rate in the Bill of Quantities for the item multiplied by the final total quantity of work done is more than 0.5% of the total of the Prices at the Contract Date.
>
> If the Defined Cost per unit of quantity is reduced, the affected rate is reduced.

This compensation event leads to a rerating of individual items where there is enough change in quantity for that item. The principle recognises that the larger the volume of work needed under a particular rate, the more efficiently (i.e. cheaply) it can usually be executed, owing to economies of scale, and vice versa. While it is the change in quantity that leads to the compensation event, the threshold at which this happens is provided financially; this must be more than 0.5% of the Total of the Prices at the Contract Date, in other words the value of the contract on day one. Bills of Quantities always contain estimated quantities when first prepared. Some of those quantities, for example structural steelwork, are unlikely to change without a change in the Scope. But others, for example earthworks, can experience significant changes from their original estimated quantities. This compensation event provides a rerating and, as explained in the final sentence, the rate can be reduced too.

Changes to rates need to reflect the contract's assessment rules, such as the schedules of cost components, the provisions for Defined Cost, Clause 52.1 and the Fee.

> **Clause 60.5**
>
> A difference between the final total quantity of work done and the quantity for an item stated in the Bill of Quantities which delays Completion or the meeting of the Condition stated for a Key Date is a compensation event.

This compensation event recognises that, with significant increases in quantity over that originally stated in the Bill of Quantities, the change in rate envisaged under Clause 60.4 will not compensate the *Contractor* for its prolongation costs. These are the time-related overheads that accrue while the works are lengthened in duration. As before, the assessment needs to stay faithful to the ECC's assessment provisions for cost and time.

> **Clause 60.6**
>
> The *Project Manager* gives an instruction to correct a mistake in the Bill of Quantities which is
>
> - a departure from the rules for item descriptions or division of the work into items in the *method of measurement* or
> - due to an ambiguity or inconsistency.
>
> Each such correction is a compensation event which may lead to reduced Prices.

The impact of this compensation event is not often well-understood in the industry. After all, in this scenario there is no additional or varied work. The issue here is that the *Contractor* has taken the Bill of Quantities at tender stage as being correct and consistent with the *method of measurement* and will have made pricing decisions based on the data provided. If an error has crept in and subsequently needs correcting by the *Project Manager*, post-contract, then either the *Contractor* or the *Client* will be compensated for the *Contractor* being misled by the incorrect documents.

This compensation event also allows a reduction in the Prices.

There is also a reference to an ambiguity or inconsistency in the Bill of Quantities, which relies on the following clause, Clause 60.7.

> **Clause 60.7**
>
> In assessing a compensation event which results from a correction of an inconsistency between the Bill of Quantities and another document, the *Contractor* is assumed to have taken the Bill of Quantities as correct.

This is one of just a few clauses in the default ECC form that establishes an order of precedence between the various components of the contract.[iv] Such instances are common but, in our experience, they rarely involve significant amounts. Any such compensation event should be administered by someone who understands the subtleties of Bills of Quantities and different methods of measurement.

61. Additional compensation events in secondary Options

> Clause X2.1
>
> A change in the law of the country in which the Site is located is a compensation event if it occurs after the Contract Date. If the effect of a compensation event which is a change in the law is to reduce the total Defined Cost, the Prices are reduced.

This compensation event occurs when a change in law happens after the Contract Date, even if the change was known about by the Parties prior to the Contract Date. The change in law must occur in the same country as the Site. So, for example, if a change in law occurs in another country where Plant and Materials are being manufactured, this will not lead to a compensation event. Tenderers bidding for an ECC contract should not therefore include for a known change in law at the Site in their proposals.

> Clause X12.3(6)
>
> The Core Group may give an instruction to the Partners to change the Partnering Information. Each such change to the Partnering Information is a compensation event which may lead to reduced Prices.
>
> Clause X12.3(7)
>
> The Core Group prepares and maintains a timetable showing the proposed timing of the contributions of the Partners. The Core Group issues a copy of the timetable to the Partners each time it is revised. The *Contractor* changes its programme if it is necessary to do so in order to comply with the revised timetable. Each such change is a compensation event which may lead to reduced Prices.

These compensation events refer to changes in multiparty collaboration arrangements. Both are self-explanatory, and the methods of assessment will reflect the main Option used. The *Contractor* will be compensated for the effect of the changes in the Core Group's arrangements, once they are known and instructed.

> **Clause X14.2**
>
> Delay in making the advanced payment in accordance with the contract is a compensation event.

Should a delay occur in the *Client* making an advance payment under secondary Option X14, where this operation is incorporated, this compensation event will occur. The payment is due under the second list item of Clause 50.3. The failure to pay that amount on time will surely lead to an obligation to pay interest under Clause 51.3. But the reality of advanced payments is often that the payment is needed as a down payment to a supplier or Subcontractor for a major piece of offsite manufacture. Without access to working capital, the *Contractor* will have relied on this payment and will suffer through its late payment. We suggest that demonstrating the impact of this event will be difficult, even considering the clear inconvenience that the late payment will cause.

> **Clause X15.2**
>
> If the *Contractor* corrects a Defect for which it is not liable under the contract it is a compensation event.

This clause is potentially contradictory, albeit its intent is clear. If there is a Defect, according to the definition in Clause 11.2(6), it is a breach by the *Contractor* and there is therefore a liability for the *Contractor* to correct. If the design has been provided with the skill and care required, then there is no Defect.

The clause is seeking to compensate the *Contractor* for rectifying issues that arise where the *Contractor's* design has been provided with the skill and care required, but where the design has not achieved the result specified elsewhere in the contract. If the *Project Manager* instructs the *Contractor* to remedy such an issue, there would be a compensation event.

> **Clause Y2.5**
>
> If the *Contractor* exercises its right under the Housing Grants, Construction and Regeneration Act 1996 as amended by the Local Democracy, Economic Development and Construction Act 2009[v] to suspend performance, it is a compensation event.

The ECC is succinct on this issue. The legislation named in Clause Y2.5 (and its equivalent in Northern Ireland) allows the *Contractor* to suspend performance in certain situations and requires contracts (for them to comply with the legislation) to contain provisions for the *Contractor* to recover its costs in those situations.[vi] That is the sole purpose of this clause.

62. Processes for compensation events

Compensation events feature in all NEC contracts and share common processes. To some users of the contract, the processes appear excessive. But they are designed to assist the Parties to deal with the consequences of change and other project risks. Where the processes are not adhered to, it is more likely that a dispute will occur. Many of the contract's requirements are for communication so that the *Contractor* and the *Project Manager* are aware of each other's actions. The processes contain several time bars, and these are to incentivise the right behaviours. There are four stages.

- **Notification**. Before committing resources to dealing with the commercial niceties of compensation events, the Parties need to establish if a compensation event has occurred. The obligations to notify compensation events are clearly set out and should be simple to observe. However, we repeatedly see disputes caused by the reluctance of *Project Managers* to notify compensation events, supposedly in the hope that the *Contractor* will not notice that something has changed or that a risk event has occurred. A *Project Manager* who omits to notify a compensation event runs the risk that the *Contractor* will do so long after the event and may still have the entitlement to do so. Dealing with such issues at the time is usually easier.
- **Quotation**. Having established that a compensation event has occurred, the first step is for the *Contractor* to prepare a quotation once instructed to do so by the *Project Manager*. Collaboration at this stage will certainly aid the avoidance of disputes. Rather than preparing a quotation in the dark, so to speak, the *Contractor* can obtain the input of the *Project Manager* before starting the quotation. There is a greater likelihood that the quotation will be accepted if it has been prepared with the *Project Manager's* knowledge. There are formal ways for the *Project Manager* to make an input, but we recommend that less-formal collaboration is also deployed.
- **Assessment**. The contract, in Clause 63, sets out the assessment rules for compensation events. These rules should be used by *Contractors* preparing quotations and by *Project Managers* undertaking their own assessments. These rules cover the impact on both time and money. They detail the *Contractor's* entitlement and the *Client's* obligations; the sole remedy clause (Clause 63.6) explains this.
- **Implementation**. This phase of the process concerns the implementation of the revised Prices, Key Dates and Completion Date(s) in the contract.[vii] It does not refer to the physical implementation of any works that are the subject of the compensation event. This is the conclusion of the process and the point after which the compensation event cannot be reassessed except by mutual consent (Clause 12.3) or the intervention of the *Senior Representatives*, the *Adjudicator* or the *tribunal*, as appropriate.

63. Notification

The purpose of the notification clauses is to ensure that both Parties are aware of the progress of change in the *works*. Managing change is difficult; if control is not maintained, the Parties will struggle to understand the true position of the project, with respect to time and money. It cannot be overstated how important this process is in the aim of avoiding disputes. The starting point of the compensation events process is the notification, where either the *Project Manager* or the *Contractor* notifies the other as to the occurrence of a compensation event. No other administration is needed at this point, such as a quotation or a programme. The first stage is to accept that there has been, or is likely to be, a compensation event.

Clause 61.1 requires the *Project Manager* to notify the *Contractor* of a compensation event that arises from the *Project Manager* or *Supervisor*

- giving an instruction (Clauses 60.1(1), 60.1(4), 60.1(7), 60.1(10) and 60.6)
- giving a notification (Clauses 60.1(17) and 60.1(20))
- issuing a certificate (Clause 60.1(15))
- changing an earlier decision (Clause 60.1(8)).

The relevant compensation events should all be known to the *Project Manager*; hence, the obligation to notify them to the *Contractor*. The notification should be provided at the same time as the communication that created the compensation event. Failure to notify a compensation event at this point is not a rare event. Some *Project Managers* have the (mistaken) belief that if they don't notify a compensation event then somehow the *Contractor* will miss the relevance of the communication and a change to the Prices may be avoided. This is unlikely, and the omission of this important contractual process is likely to cause unnecessary dispute.

Where the *Project Manager* notifies a compensation event to the *Contractor* in this manner, the notification must be accompanied by an instruction to submit a quotation. Potentially, therefore, we may have three items of communication covering the same event

- the original instruction, certificate or decision
- a notification of a compensation event
- an instruction to submit a quotation (although this can be included in the notification).

In Clause 61.2 (and Clause 61.4), the contract allows for the *Project Manager* and the *Contractor* to agree that there is a compensation event but for the *Project Manager* to state that there is no impact on time and money. While such a provision may seem like a futile exercise in contract administration, it does leave open the door to a revision to the latter opinion through adjudication.

Clause 61.3 describes the process to be followed when the *Contractor* is obliged to notify a compensation event and when the *Project Manager* has omitted to notify an event that should have been notified by the *Project Manager*.

The important part of this clause is the time bar of eight weeks to the *Contractor's* notification of a compensation event when that event should have been notified by the *Contractor*. Failure, by the *Contractor*, to notify a compensation event within the stated period means that 'the Prices, the Completion Date or a Key Date are not changed'.

This clause is frequently amended by a Z clause reducing the time-barred period from eight weeks to a much shorter period. The shortest period that we have seen is five working days, which appears to be far too short. The purpose of this time bar is to encourage the *Contractor* to make the *Project Manager* aware of a compensation event when otherwise the *Project Manager* might not be aware. The obligation is not to make the *Project Manager* aware (that's far too subjective and difficult to demonstrate) but to notify, remembering the provisions of clauses 13.1, 13.2 and 13.7. Legal commentators, better qualified than us, point to various uncertainties with this clause, particularly where it is used as a limit or exclusion of liability by the *Client*. We will avoid joining that discussion, but where Z clause amendments are being prepared, they need to be done by someone with the necessary knowledge of the wider legal issues involved. This clause has a constructive purpose, which is to encourage collaboration. If the time period is set so short as to be impractical, it may affect the legal enforceability of the clause and will certainly make the Parties' correct use of the contract more challenging. In both scenarios, the creation of a dispute is more likely.

We have the following advice for those drafting notification provisions in contracts.

- Don't alter the eight-week period, unless there is a good operational reason for doing so.
- If the eight-week period is to be altered, consideration should be given to how the revised period will work in practice.
- Don't try to improve the drafting around the notification periods (for example, the reference to the *Contractor's* beliefs in Clause 61.3 is often amended).
- As with all amendments, only appoint someone to draft these who has the necessary skills and qualification to do so.

Our advice to those managing notifications in contracts is as follows.

- Understand the combined effect of the various compensation event clauses; core clauses, secondary Options and Z clauses.
- View the time bars as a maximum and make your notifications as early as you can.
- Adhere to the formalities in Clause 13 with respect to the format of communications.

There are two possible responses by the *Project Manager* to a compensation event notification.

- Notify the *Contractor* that, for one or more of the reasons stated in Clause 61.4, the Prices and dates will not be changed.
- Notify an instruction to the *Contractor* to submit a quotation(s) for the event.

When the *Project Manager* instructs the *Contractor* to submit a quotation, the following statements can be made by the *Project Manager*.

- That the *Contractor* did not give an early warning of the event which an experienced contractor could have given (Clause 61.5). See our discussion of this issue in Chapter 3.
- Assumptions as to the assessment of the compensation event where there is uncertainty (Clause 61.6). Where those assumptions are subsequently corrected, a further compensation event will occur under Clause 60.1(17).

The final part of Clause 61 states that a compensation event cannot be notified after the issue of the Defects Certificate. The Defects Certificate is issued by the *Supervisor* on or shortly after the *defects date* and, coming long after Completion, is an action that is often delayed or even missed entirely. This is one of several connections to that clause that reinforce the need for it to be complied with.

64. Quotation

The *Contractor's* quotation must follow several rules in the contract, not all of which are immediately apparent from Clause 62.

Clause 62.1 allows the *Project Manager*, following discussions with the *Contractor*, to instruct the submission of alternative quotations. The *Contractor* must comply with this instruction, but may also provide further quotations based on alternative methods. For example, an instruction for additional work may be satisfied by the use of prefabricated components, or alternatively by in situ ones. The *Project Manager* could instruct the *Contractor* to provide quotations for both methods. The working of this clause sits well with Clause 61.6, explained earlier.

Clauses 62.2 and 62.3 requires that a quotation should be submitted by the *Contractor* within three weeks of instruction and must include

- proposed changes to the Prices
- any delay to the Completion Date
- any delay to the Key Date(s).

These terms are all defined and have a specific meaning in the contract. We routinely see quotations from *Contractors* (and indeed assessments prepared by *Project Managers*) that do not follow the contract. The use of the term 'Prices' is specific to the main Option selected. The cost and time impact of a compensation event should be assessed in accordance with the provisions in Clauses 63.1 and 63.3, respectively; the terms in those clauses are also defined and some of them are also main-Option specific.

The quotations should contain enough explanation of the assessments and should also include any necessary alterations to the Accepted Programme. Where the compensation event relates to the Information Execution Plan (Clause X10.5), any changes to that must also be provided in the quotation.

We describe in a moment what the *Project Manager* must do once a quotation is received. Differences of opinion are predictable in this type of arrangement, but those differences will be exacerbated if the quotation (and any response from the *Project Manager*) is not prepared in accordance with the contract. Where the quotation has been prepared in accordance with the contract, any differences of opinion can be addressed much more easily.

The *Project Manager's* response to a quotation is one of the following.

- Accept the quotation, in which case the compensation event is implemented on the terms in the quotation.
- Instruct the *Contractor* to submit a revised quotation, having explained the problem(s) with the previous quotation as required in Clause 62.4.
- Tell the *Contractor* that the *Project Manager* will now assess the compensation event.

If the *Project Manager* elects to make the assessment, the procedure is explained in Clause 64, which we describe later.

65. Assessment

The assessment of a compensation event, either by the *Contractor* in a quotation or in a *Project Manager's* assessment, must comply with the contract. We explained earlier how a compliant structure to the assessment will make the resolution of any differences much easier. The two most important subclauses in Clause 63 are Clauses 63.1 and 63.5, as they are the governing provisions for the cost and time impacts, respectively, of compensation events. They are assessed simultaneously; the contract does not recognise the separation of these two issues, although we do often see parties seeking to generate discrete compensation events in that manner.

The default method for assessing the financial impact is in Clause 63.1, requiring the Defined Cost and Fee to be assessed.

> **Clause 63.1**
>
> The change to the Prices is assessed as the effect of the compensation event upon
>
> - the actual Defined Cost of the work done by the dividing date,
> - the forecast Defined Cost of the work not done by the dividing date and
> - the resulting Fee.
>
> For a compensation event that arises from the *Project Manager* or the *Supervisor* giving an instruction or notification, issuing a certificate or changing an earlier decision, the dividing date is the date of that communication.
>
> For other compensation events, the dividing date is the date of the notification of the compensation event.

Calculation of the Fee is straightforward once the Defined Cost has been established. The term 'Defined Cost' is a defined term and its meaning is particular to each main Option. In main Options A and B, it should refer to the Short Schedule of Cost Components; in main Options C, D and E, it should refer to the Schedule of Cost Components. So, an immediate test of the compliance of a quotation or assessment is whether its structure matches the eight headings in the schedules. If so, read on. If not, ask the originator to start again.

A common area of dispute in assessing compensation events is the difference between the first and second list items of Clause 63.1. Many people are uncomfortable with the subjectivity of forecasting future costs and prefer to delay the assessment, monitor actual costs and apply a wholly retrospective assessment. Not only does that go against what the contract states, but it potentially skews the payments in favour of one of the Parties. We are aware of some case law involving an NEC3 contract that supports the fully retrospective approach; to be frank, neither of us agrees with the decision of the court on that occasion. Abiding by the contract's provisions is not difficult and that should always be the default approach.

The *Project Manager* and the *Contractor* may agree instead to use 'rates or lump sums' instead of this methodology (Clause 63.2); the contract is not specific about where those rates and lump sums come from. One possibility is the existing Bill of Quantities or Activity Schedule (as appropriate); another might be industry publications. This route requires agreement; therefore, neither Party should be in a position where it is forced to accept rates that it isn't comfortable with.

Some, but not all, compensation events lead to a reduction in the Prices; this is stated in the relevant clause. Clauses 63.3 and 63.4 provide for this process. There is unlikely to be any dispute in these circumstances. The contract is clear as to which compensation events may lead to reduced Prices.

We mentioned the importance of Clause 63.5 earlier. This clause provides for the assessment of the time-related impact of a compensation event. The impact on the Completion Date, sectional Completion and Key Dates is established via this rule.[viii]

> **Clause 63.5**
>
> A delay to the Completion Date is assessed as the length of time that, due to the compensation event, planned Completion is later than planned Completion as shown on the Accepted Programme current at the dividing date.
>
> A delay to a Key Date is assessed as the length of time that, due to the compensation event, the planned date when the Condition stated for a Key Date will be met is later than the date shown on the Accepted Programme current at the dividing date.

The drafters added in a sentence to Clause 63.5 not found in previous editions. This is shown in the box below. This did not seem to get universal appreciation, although users, through social media boards in particular, were asking for some clear words to spell out whether or not the Accepted Programme should be used without change in assessing compensations or, if it should be changed, then for what.

> **Clause 63.5, June 2017**
>
> When assessing delay, only those operations which the *Contractor* has not completed, and which are affected by the compensation event, are changed.

In making some minor changes to typos, the drafters took the opportunity to lay down the specific instances that should be taken into account in what you do with the correct version of the Accepted Programme in assessing each compensation event. Time will tell as to whether this helps users to process this part of the contract with greater certainty. The authors believe this will add to the clarity but does not materially change what we considered to be the correct approach anyway, without being so explicitly stated before.

> **Clause 63,5, January 2019**
>
> The assessment takes into account
> - any delay caused by the compensation event already in the Accepted Programme and
> - events which have happened between the date of the Accepted Programme and the dividing date.

The dividing date is introduced in Clause 63.1. The importance of the Accepted Programme can be seen from this clause. The *Contractor's* time entitlement in a compensation event always refers to the Accepted Programme. If that is a product of an efficient process, the task of establishing the time impact should be easier than the alternative scenario. Quotations and assessments need to contain an explanation of how they have been prepared. Clause 62.2 requires a quotation to contain any alterations to the Accepted Programme.

The issue of establishing the additional time in any form of construction contract can be difficult, whatever form is used. But the NEC processes depend on the programme actions being kept up to date and used as a management tool. Too often we see adjudications where, for differing reasons, the programme has not been kept up to date and that generates disputes where none or fewer should exist. We have encountered *Contractors* who 'don't do programmes' and *Project Managers* who avoid accepting programmes with the very aim of torpedoing the compensation event process. Both approaches will provide the torpedo. Those who avoid their own programme obligations are kidding themselves if they expect anything other than difficulty when compensation events are assessed. Even where the programme is maintained properly, the assessment of the impact of the compensation event in accordance with Clause 63.5 can be difficult and is a routine part of dispute resolution.

We have already described the impact of the *Contractor* not giving an early warning that could have been given. If the *Project Manager* has stated that this is the case, the event will be assessed as if an early warning had been given. We provided more discussion on this subject in Chapter 3. For Clause 63.7 to apply and for the assessment to be affected, the *Project Manager* must have made the statement required in Clause 61.5.

Clause 63.8 provides a challenge for those quoting and assessing compensation events.

> **Clause 63.8**
>
> The assessment of the effect of a compensation event includes risk allowances for cost and time for matters which have a significant chance of occurring and are not compensation events.

The implication of this clause is that there must be a risk to the *Contractor*, so the allowances must be for a risk that isn't borne by the *Client*. This clause is subjective in its application so tends to generate some debate between the Parties. It only applies to matters that have a 'significant' chance of occurring. The word 'significant' isn't defined, nor does the contract state how the *Contractor* should allow for risks with an insignificant chance of occurring. Hence, the addition of allowances, both financial and time, causes difficulties for the Parties. While the addition of risk allowances alone would not be likely to cause a dispute, this clause features regularly in NEC adjudications. Not all participants apply it to time issues, seemingly unaware that it expressly includes time.

There is a sensible restriction on the assessment of time and money in Clause 63.9; anything assessed must be reasonably incurred and not incurred due to the *Contractor's* delayed response to an instruction.

Clause 63.10 reflects the position in law in several jurisdictions. Responsibility, and hence liability, for errors in the contract rest with the Party that provided the Scope. The wording is odd in one respect, in that the ambiguity or inconsistency is not expressly stated as being in the Scope, but the clause only applies when there is a change to the Scope.

Where assessments of compensation events are undertaken in a contract that includes Clause X1 (price adjustment for inflation), there are additional assessment rules to consider in Clause X1.5, when considering Defined Cost. This is straightforward arithmetic and should not cause undue difficulty. Once again, the dividing date in Clause 63.1 is used.

66. The *Project Manager's* assessments

We have already seen what the *Project Manager* must do once a quotation has been received from the *Contractor*. The references in Clause 62.3 are added to by Clauses 64.1 and 64.2.

In Clause 64.1, the *Project Manager's* role in assessing compensation events is set out for the situation where

- the *Contractor* has not submitted the quotation within the time allowed
- the *Project Manager* decides that the quotation has not been prepared correctly
- the quotation did not include the programme information required
- at the time of submitting the quotation, the *Contractor's* latest programme has not been accepted for a reason stated in the contract.

Clause 64.2 extends the *Project Manager's* authority in this area by requiring the programme to be assessed by the *Project Manager* for remaining work and then used in the assessment of the compensation event, where

- there is no Accepted Programme
- the *Contractor* has not submitted programme information required by the contract
- the *Project Manager* has not accepted the *Contractor's* latest programme for one of the reasons in the contract.

All seven of these situations lead to the *Contractor* losing some or all input into the assessment process. Each cause reflects the apparent failure of the *Contractor* to comply with the contract; therefore, these are matters to which the *Contractor* should dedicate attention. But several of these situations must be judged by the *Project Manager* and it is common to hear disputes about those judgements even before the assessment begins. Where the *Project Manager* decides to accept or not accept something, the implications of the decision must be properly considered. Some of the situations in our lists are a matter of fact; for example, the *Contractor* has either responded within the time period or not. Other items, such as the decision that a quotation has not been correctly prepared, by way of example, can be subjective.

Where the *Project Manager's* need to assess a compensation event becomes 'apparent', the *Contractor* must be notified and the *Project Manager* has the same time period that the *Contractor* had (Clause 64.3). Normally, that will be three weeks (Clause 62.3), but that may have been lengthened by agreement (Clause 62.5).

67. Proposed instructions

We explained the rationale behind Clause 65 earlier, in our commentary on Clause 60.1(20), which is linked to this clause. Where the *Project Manager* instructs the *Contractor* to submit a quotation, this must be done, but without undertaking the work in question. The normal period of three weeks for the *Contractor* applies here. The *Project Manager* must state the period during which an instruction to proceed with the works will be given when instructing the quotation. The drafting of the instructions to proceed and the explanation of when the work is likely to take place are important considerations. The better that this is drafted, the more likely it is that the *Contractor* will understand what is required and, at least in theory, the more compliant the quotation is likely to be. Where the *Project Manager's* instructions are vague or inconsistent, the process is unlikely to start well.

The *Project Manager* has some familiar choices when making a response to the quotation.

- Instruct the *Contractor* to submit a revised quotation, presumably having explained the problem with the previous quotation.
- Accept the quotation, instruct the works and notify a compensation event. There are three communications here, so the *Project Manager* should issue them separately.
- Notify the *Contractor* that the quotation is not accepted.

This final item then triggers a compensation event under Clause 60.1(20), discussed earlier.

Clause 65.3 allows the *Project Manager*, if an acceptable quotation cannot be obtained through this route, to instruct the change (presumably to the Scope) through the other routes in the contract, notify a compensation event and instruct the *Contractor* to submit a further quotation. This then allows the *Project Manager* to assess the compensation event if the next quotation from the *Contractor* is still unacceptable. This process is unduly burdensome and justifies the need for the *Project Manager* and the *Contractor* to meet and discuss these issues before committing to a process that could take many weeks to conclude.

Contractors may be comforted by the compensation event in Clause 60.1(20), but it is unlikely to placate those who complain of being used as a free estimating service. It is unlikely that the consequence of this process will have any impact on the Completion Date or Key Dates, so the likely discussion area will be that of Defined Cost. Where the unaccepted quotation has been prepared by a member of the *Contractor's* project team, it might be difficult to demonstrate any impact on Defined Cost, particularly in an environment where the person in question is paid a fixed salary with no payments for overtime. The *Project Manager* in this scenario may argue that no additional cost has been incurred by the *Contractor*. Some *Contractors* now outsource this work to external providers as a way of demonstrating and recovering the costs involved.

Where unaccepted quotations involve design and build work, the costs may be more significant than with build-only projects. Keeping records of the costs of, say, designers, planners and estimators is vital. Compensation events under Clause 60.1(20) are likely to involve the assessment of the actual Defined Cost of work already done, rather than the forecast of work to be done. Even with good records, there are likely to be areas for discussion as to the time and cost commitment made by the people involved. In main Option C, D, E and F contracts, the *Project Manager* can rely on the drafting in Clause 52.2 to verify information, but there are no such provisions in contracts formed with main Options A or B. This process may be drawn out if the *Contractor* has worked with Subcontractors also employed under the same provisions, who will likewise wish to recover their own costs.

68. Implementation

Compensation events are implemented at the conclusion of one of the processes in the contract.

- The *Project Manager* accepts the *Contractor's* quotation (Clause 62.3). This, incidentally, is the line of least resistance!
- The *Project Manager* notifies the *Contractor* of an assessment made by the *Project Manager* (Clause 64.3).
- The *Contractor's* quotation is treated as having been accepted (Clauses 62.6 and 64.4).

Once implemented, a compensation event cannot be revisited, except by the dispute resolution process (Clause 66.3). If assumptions had previously been stated by the *Project Manager* under Clause 61.6, any correction to these under Clause 60.1(17) will not revisit the original compensation event but will require a further compensation event to be processed in order to address the correction. Where one of the Parties is unhappy with the implementation terms, it needs to establish what it wishes to do. Its options are limited; try to negotiate with the other Party or refer to the selected dispute resolution process. There really isn't an alternative that avoids a prolonged exchange of communications going over old ground for months to come.

Clause 65 does not expressly state that the *Project Manager* should notify the *Contractor* and *Client* of the final implemented terms, but we think it is good practice for this to happen to assist with later audit and verification activities.

69. Time restrictions on the *Project Manager* and extending time

We saw earlier that the *Contractor* has a time bar of eight weeks (or fewer, if amended by a Z clause), within which it must notify many of the compensation events in Clause 60.1 and elsewhere.

The *Project Manager* also faces time bars in relation to failures to respond to compensation event notifications, in three situations where a response has not been provided on time by the *Project Manager*.

- The *Contractor* has notified a compensation event (Clause 61.4).
- The *Project Manager* has stated that it will assess a compensation event but hasn't done so by the date specified in the contract (Clause 64.4).
- The *Contractor* has submitted a quotation (Clause 62.6).

In all three situations, the *Contractor* may notify the *Project Manager* of the failure to respond within the respective time period (emphasis added). Two weeks after that notification, the original communication is treated as being accepted. Note that the ECC doesn't use the term 'deemed'.

The *Project Manager* should not get caught out like this. Adherence to the original timescales in each of these three situations should be possible and, with a reminder from the *Contractor*, there is little excuse for the *Project Manager's* failure. Correspondence of this nature needs to be clear. While the contract does not expressly require the quotation of a clause number, we suggest that the *Contractor's* reminder should do so and should be very clear as to what it is saying. If the *Contractor* wishes to rely on that reminder in later adjudication or at the *tribunal*, this record, as opposed to a loosely worded communication, will assist in demonstrating that the *Project Manager* was properly reminded.

The time periods in the compensation clauses are generally fixed and each situation is unique. Therefore, we can expect quicker responses in some situations and slower responses in others. The contract allows for the extension of time periods

- for the *Project Manager* to reply to the *Contractor's* notification of a compensation event (Clause 61.4)
- for the *Contractor* to submit a quotation (Clause 62.5)
- for the *Project Manager* to reply to the quotation (Clause 62.5).

Extending the time allowed under Clause 62.5 requires the additional time to be agreed by the *Project Manager* and the *Contractor* before the original time period has expired. If you must extend the time allowed for, say, a three-week period, then don't wait until day 20 to do so. You are far more likely to gain the consent of someone else if you ask early on. Remember when asked to consent to an extension that you may be the one asking next time, so agree to more time if the programme allows.

70. Dealing with compensation events in adjudication and at the *tribunal*

Where the Parties cannot agree on the assessment of compensation events then adjudication, and possibly the *tribunal*, will become the next options. Of course, there is the possibility of mediation and the role of the *Senior Representatives*, which we describe later in this book.

Where a Party wants to dispute the assessment of a compensation event it may do so.

- **Clause W1.1(4)**. Either Party may dispute a decision of the *Project Manager*; it may refer the dispute to the *Adjudicator* within four weeks of the Party becoming aware of the decision. This phrase seems difficult to establish. When did someone become aware? That's possibly difficult to establish, so the referring Party might need to establish, through a document audit trail, when the responding Party became aware. We think that this clause would be easier if it referred to when the referring Party had been notified of the compensation event. But this clause refers to a wider set of issues, other than just compensation events, hence the looser wording.
- **Clause W1.1(4)**. The *Client* can dispute the circumstances of a compensation event communication being treated as accepted. This must be done within four weeks of the communication being treated as accepted.
- **Clauses W1.3(7) and W2.3(7)**. The *Adjudicator* must assess issues of cost and time caused to the *Contractor* in the same way as a compensation event is assessed. In other words, the *Adjudicator* must act in the same way as the *Contractor* and *Project Manager* should have acted earlier.
- **Clauses W1.4(2) and W2.4(2)**. Where the *Adjudicator's* decision still leaves a Party dissatisfied, that Party must issue a notice of dissatisfaction to the other Party within four weeks of being informed of the *Adjudicator's* decision.
- The *tribunal* then deals with the dispute that is referred to it, which may be the entirety of the *Adjudicator's* decision or just a part of it.

In disputing a compensation event assessment, Parties should consider the following.

- Have the negotiation options and the *Senior Representatives* route been exhausted?
- Set the Notice of Adjudication carefully and define the *Adjudicator's* jurisdiction precisely. In essence, set the *Adjudicator* a question that (a) can be answered and (b) is likely to assist your contract position.
- When referring a dispute to the *tribunal*, review how the adjudication went and potentially amend the case before asking the *tribunal* to consider it. Clauses W1.4(4) and W2.4(3) allow this amendment of the case to take place. The dispute referred to the *tribunal* cannot change from the one decided by the *Adjudicator*, but the way it is described may be.

NOTES
i By agreement of the Parties in Clause W2.
ii Clause 25.2.
iii The ECC sets a precise test in Clause 60.1(13), which we have paraphrased here.
iv See also Clauses 60.1(1), 60.3 and 63.10.
v HMG (Her Majesty's Government) (1996) Housing Grants, Construction and Regeneration Act 1996. The Stationery Office, London, UK. Amended by HMG (2009) Local Democracy, Economic Development and Construction Act 2009. The Stationery Office, London, UK.
vi As defined in section 112(3A) of the Housing Grants, Construction and Regeneration Act 1996, as amended, and in article 11(3A) of the Construction Contracts Order (Northern Ireland) 1997, as amended.
vii There may be more than one if secondary Option X5, sectional Completion, forms part of the contract.
viii See Clause X5.1.

Chapter 7
Termination

71. Termination of contracts generally

We explain later how the standard position for termination in NEC contracts is not to terminate the contract but to terminate the further involvement of the supplier in question. All legal jurisdictions, or at least those likely to be of interest to NEC contract users, have provision for the termination, properly or otherwise, of contracts and for the consequences of termination to be dealt with after the event. This could spawn a book all on its own and we do not profess to have the requisite knowledge to address such a broad subject. Therefore, our explanation of the termination provisions here is confined solely to the ECC and ignores entirely the other termination options available to a contracting party.

We end this section on termination with a warning to the Parties. Termination can be a contractual minefield. If you get the contractual termination process wrong in an NEC contract (or indeed any other) then you will be required to deal with the consequences provided for by the legal jurisdiction in question. That will almost certainly prove to be expensive and time-consuming. So, as with our advice in almost every other chapter of this book, we recommend that the termination provisions are followed carefully. Unlike our advice in many other parts of this book, we also recommend that a terminating Party takes appropriate legal advice before starting the termination.

72. Termination in the NEC4 ECC

There are several places to look in the ECC for termination provisions

- Payment (normal assessment, on termination) – Clause 50.1
- Payment (final assessment, on termination) – Clause 53.1
- Loss or damage after termination – Clause 80.1
- Section 9 of the core clauses and the appropriate main Option
- Termination by the *Client* – Clause X11
- Effect on Project Bank Account – Clause Y1.14
- Payment notices on termination – Clause Y2.4.

We have described the termination provisions in a logical order, which doesn't necessarily follow the clause numbering and layout of the contract.

73. What is being terminated?

The termination provisions begin at Clause 90.1.

> Clause 90.1 (excerpt)
>
> If either Party wishes to terminate the *Contractor's* obligation to Provide the Works it notifies the *Project Manager* and the other Party giving details of the reason for terminating […]

So, the contract is not being terminated. The *Contractor's* obligation to Provide the Works is being terminated. This is part of the obligation in Clause 20.1, 'the *Contractor* Provides the Works in accordance with the Scope'. The omission of the final five words of this obligation probably means little. The interpretation of Clause 90.1, therefore, suggests that all other provisions of the contract remain in place after termination.

74. Termination for convenience

The NEC drafters made a subtle change to the termination provisions when NEC4 was published. The long-standing ability of the *Client* to terminate for any reason has been moved from the core clauses to a secondary Option clause, Clause X11.

> Clause X11 (excerpt)
>
> The *Client* may terminate the *Contractor's* obligation to Provide the Works for a reason not identified in the Termination Table by notifying the *Project Manager* and the *Contractor*.

This change may seem minor. Indeed, we anticipate that most clients, when drawing up documents at tender stage, will include secondary Option Clause X11 in the list at the start of Contract Data part one. But clients now have the choice whether to do so and this will send a message to tenderers, along with other provisions of the contract, as to how this client intends to conduct itself during the *works*. As we will see later, the reason for termination has an impact on the payments made following termination.

75. Starting the termination

The termination process starts with one of the Parties notifying the other Party and the *Project Manager* of its intention, as we saw earlier in the excerpt quoted. Clause 90.1 continues.

> Clause 90.1 (excerpt)
>
> [...] The *Project Manager* issues a termination certificate promptly if the reason complies with the contract.

We consider the *Project Manager's* role later, but we must highlight the difficult task that the *Project Manager* faces here in establishing whether the reason complies with the contract.

Clause 90.2 introduces the Termination Table. The contract capitalises the term but it doesn't seem to be defined anywhere, certainly not in Clause 11.2 with other defined terms. Clause 90.2 and the Termination Table allocates the appropriate combination of procedures (P1, P2, P3 or P4) and payment amounts (A1, A2, A3 or A4) to the reason (R1, R2, R3 etc.) provided for termination. Where secondary Option Clause X11 has been included in the contract, additional statements are included therein for the Termination Table.

Clause 90.3 requires that the procedures for termination are implemented immediately after the *Project Manager* has issued the termination certificate. So, the certificate is a condition precedent for the procedures to commence. Remember that the *Project Manager* must issue the certificate 'promptly'. Any delay in that process will delay the termination procedures and, indeed, delay the cessation of the *Contractor's* work as stated in Clause 90.4.

76. The reasons for termination

The termination provisions rely on a set of reasons, referred to as R1 to R22, for the correct allocation of procedures and payments. It isn't necessary to list them verbatim here, but we have provided a summary in Table 7.1.

Table 7.1 Summary of reasons for termination

R1 to R10	Insolvency	Both Parties
R11 to R15	*Contractor* default	Termination by *Client*
R16	Non-payment	Termination by *Contractor*
R17	Release	Both Parties
R18 to R20	Failure to restart work	Both Parties
R21	Force majeure	Termination by *Client*
R22	Corrupt Act	Termination by *Client*[i]

On the face of things, these reasons may appear straightforward. But in detail many of them are not. The 'force majeure' reason, R21, by way of example, relies on the subjective judgement of three people (*Client*, *Contractor* and *Project Manager*) to establish whether the event in questions qualifies as a reason for termination. How small a chance are we considering? Insolvency of a Subcontractor? Brexit? An Icelandic ash cloud? In a termination scenario, the Parties are always going to view things differently.

Of particular note are the first ten reasons, those referring to insolvency events. Having to establish that the reason complies with the contract' is a challenging task for anyone without professional knowledge of insolvency. In some legal jurisdictions, the words used in R1 to R10 may have specific meanings in law. Typically, in construction, the people taking on *Project Manager* roles may be quantity surveyors, engineers, architects or tradespeople. None of these types of people will have the necessary professional skill to judge insolvency issues. Therefore, where a reason is quoted by the terminating Party, the *Project Manager* must obtain assistance to check the reason if the necessary skills aren't immediately available.

How does one substantially hinder the *Client* or substantially break a health or safety regulation? These are two further examples of the difficulty faced by the *Project Manager* when establishing if the terminating Party is correct.

77. The *Project Manager's* role in termination

The *Project Manager* has a role in the termination process, notwithstanding the fact that it is the Parties that are involved in termination. The *Project Manager* is the recipient of the documents that start the process, as we described earlier. But the role is much broader than that of issuing a certificate. The *Client* and the *Project Manager* will probably co-operate closely in any termination scenario where the *Client* is terminating; indeed, R11 to R15 require the *Project Manager* to make a notification of the *Contractor's*

103

default. Although, interestingly, Clauses 91.2 and 91.3 do not state who should be notified by the *Project Manager* in this regard.

78. Procedures on termination

Clause 92 sets out the four procedures that can be used on termination. The Termination Table in Clause 90 dictates which procedure should apply in any event. Clearly, much will depend on the circumstances of the termination. For example, a *Client* that is terminating due to a decision to discontinue the project will not want to use any of the *Contractor's* Equipment or assign the benefit of the subcontracts. But a *Contractor* terminating because the *Client* is insolvent will want to ensure that it recovers all its property from the Working Areas as soon as possible, perhaps in the absence of any meaningful input from the *Client* and *Project Manager*.

As with other aspects of termination, the words in Clause 92 sound deceptively simple, but their application will take tact and pragmatism.

Procedure P1 allows the *Client* to complete the *works* and use any Plant and Materials to which it has title. The issue of title is covered in Clauses 70.1 and 70.2.

> Clause 70.2 (excerpt)
>
> Whatever title the *Contractor* has to Plant and Materials passes to the *Client* if they have been brought within the Working Areas.

The first part of P1 allows the *Client* to complete the *works*. In some legal jurisdictions, this needs to be stated and agreed, particularly where the *Contractor* is willing and able to complete the *works* if termination didn't occur. The second part, access to Plant and Materials, presents the *Project Manager* with another judgement. In many termination situations, little Plant and few Materials may be left on the Site, owing to, say, the *Contractor's* inabilities to meet legal obligations. The past suppliers of Plant and Materials may remain unpaid from prior supply. How does the *Project Manager* establish title in this situation? Following the contract, it should be simple. Anything that is within the Working Areas should be the *Client's* title, surely? No, because title cannot be passed from *Contractor* to *Client* if the *Contractor* doesn't have clear title in the first place. After the insolvency of a *Contractor*, its former suppliers may seek to recover Plant and Materials previously supplied if that is physically possible. Recovery of ready-mix concrete is impossible for obvious reasons, but recovery of unfixed, still-boxed, air conditioning units would be straightforward. So, the *Client* and the *Project Manager* need to work closely together to establish title. External organisations are unlikely to recognise the *Project Manager's* role. Where Plant and Materials are generic, there is less programme and cost risk associated with this point. Where they are bespoke, for example structural steel and precast concrete, they may be key components and replacement will take time. Disputes with the *Contractor* and its suppliers must be managed and resolved where the *Client* intends to complete the project and benefit from the supply relationships previously established by the *Contractor*.

Procedure P2 assumes that the *Contractor* will remove everything from the Site and assign the benefit of any supply contract to the *Client*. The assignment of a subcontract is something that must be provided for in advance when a subcontract or supply contract is established. Presumably, because of the wording in Clause 92.1, the Plant and Materials that must be removed from the Site by the *Contractor* under P2 are those to which it doesn't have title, which it therefore cannot transfer to the *Client*.

Procedure P3 allows the *Client* to use Equipment to which the *Contractor* has title to complete the works. This could be crucial, for example as the continued use of tower cranes or bespoke steel formwork for concrete structures. The removal and replacement of such items would be time-consuming and no doubt expensive. Repossession of such items will, in practice, prove difficult if title doesn't reside with the *Contractor*. So, again, the *Client* must manage relationships with third parties, and hopefully avoid entering costly or time-consuming disputes with them.

Procedure P4 requires the *Contractor* to leave the Working Areas and remove Equipment. That is a succinct statement and one unlikely to lead to any debate about carrying it out.

The Termination Table sets out which procedures apply based on which reason has been relied on. Procedure P1 applies in all five situations in Clauses 90.2 and X11.2. The remainder apply in certain situations, depending on the reason, which itself relates to the event that led to termination.

79. Payments due on termination

The payments due on termination are labelled with the letter A, for 'amount'. As with procedures, the selection of payment(s) due is related to the reason for termination.

Amount A1 is due in all situations; it is almost like a 'settling up' of the various amounts due under the contract

- 'normal' payments
- Defined Cost of Plant and Materials to which the *Client* has title and possession and of which the *Contractor* is obliged to accept delivery (i.e. cannot opt out of purchasing)
- Defined Cost reasonably incurred by the *Contractor*
- any amounts retained by the *Client* (e.g. Clauses 50.5 and X16) (the return of retention under Clause X16 may trouble the *Client* when the longer-term position on Defects might not be clear, but the contract is clear that this should be done)
- any unpaid balance of an advance payment (Clauses X14).

Amount A2, where appropriate, is paid to remove the Equipment from the Working Areas. It is a forecast of the Defined Cost of doing so. This amount will only be payable by the *Client* if the termination has not resulted from a default of the *Contractor*.

Amount A3, where appropriate, is a deduction of the amount forecasted to be the additional amount to be paid by the *Client* in completing the *works*. This amount will only be payable by the *Contractor* if the termination resulted from a default of the *Contractor*.

Amount A4, where appropriate, is a compensatory payment made by the *Client* to the *Contractor* by applying the *fee percentage* applied to the estimated value of the remaining work that will not now be undertaken by the *Contractor*.

Termination features in several other clauses in the contract.

Clause 50.1	The final assessment date will be when the *Project Manager* issues a termination certificate.
Clause 53.1	The *Project Manager* certifies a final payment no later than 13 weeks after the issue of a termination certificate.
Clause Y1.14	No further payments are made into the Project Bank Account after the issue of a termination certificate.
Clause Y2.4	Requirements for 'pay less' notices after the issue of a termination certificate.

80. Where do disputes arise in termination?

The simple answer to this question is in every aspect of the process. Some of the processes that we have described earlier are clear as to what must be done and when. But phrases such as 'the forecast Defined Cost'[ii] bring potential for disagreement. Our explanation of the difficulties in establishing that insolvency events have taken place shows that even establishing the reasons for termination may be troublesome. Assigning the benefit of subcontracts, demonstrating title and removing Plant and Materials and Equipment are all activities that require some element of co-operation. What happens if one Party won't co-operate? Or can't, owing to insolvency? These types of problem are fertile breeding grounds for disputes. Where the Parties know that their current relationship is likely to end very soon, there is little incentive to continue collaborating, unless mutual work opportunities elsewhere exist.

Our repeated advice throughout this book is to do what the contract says as a route to avoiding disputes. Termination disputes are probably the least avoidable in the industry and it will often be impossible for one Party to unilaterally prevent a dispute from happening. Referring or responding to an adjudication or litigation will be easier if the contract has been followed and if the records to demonstrate this have been retained.

NOTES

i The implication here is that, at least under the contract, the *Client* is free to commit a Corrupt Act without a contractual sanction being available to the *Contractor*.
ii Clause 93.2.

Chapter 8
NEC4 dispute procedures

81. Common dispute resolution procedures?

There is an array of different dispute procedures, available across the NEC4 contracts; however, not all the various procedures are found in all the NEC4 contracts. The drafters have tried carefully to match the dispute procedures with the particular NEC4 contract. So, for example, there are no express provisions for dealing with any disputes arising within the Dispute Resolution Service Contract (DRSC), as how many times do you expect a dispute to occur with the person(s) appointed to resolve a dispute? Where a dispute does occur, it is likely to be part of wider proceedings involving the Parties and the *Adjudicator*. We see this occasionally in adjudication enforcement cases in the courts. Also, in many jurisdictions, any dispute with an independent body is likely to be heavily coloured by local laws and not subject to any contractual agreement. It makes sense for the DRSC to have no express dispute resolution provisions, but note that the Framework Contract similarly has no express provisions, instead relying on the dispute procedure from the contract(s) used for the Time Charge Order or Work Order.

We have seen Parties add dispute procedures to the Framework Contract, and there has even been litigation in England over whether the dispute provisions of the Framework Contract or the underlying Supply Short Contract (SSC) should apply to the resolution of the dispute.[i] In that case, the two dispute provisions differed and at least one of the Parties considered that a tactical advantage could be obtained by reliance on one of the procedures at the expense of the other. The limited commitments that the Parties make to one another in the Framework Contract suggest that it is not necessary to do this. The Parties will often do this to centrally control the direction of disputes as best they can, generally through some sort of dispute escalation process. Where the Framework Contract is used, it is typically amended heavily with Z clauses to add obligations to the Parties. Once those obligations are sufficiently material, there is a possibility of dispute and hence the need for resolution procedures.

In the short NEC4 contracts, there is adjudication followed by a *tribunal*, and in most of the main NEC4 contracts there is reference to the *Senior Representatives* followed by adjudication and then a *tribunal*. In the ECC, the option to use a Dispute Avoidance Board is also available. The exception is the Alliance Contract (ALC), which offers reference to an independent expert or to the *Senior Representatives*; if the latter is chosen, a mediator may also be appointed. Adjudication or a *tribunal* are not available in the ALC, in keeping with the co-operative style of contracting. This seems to provide a difficulty in the UK, with the statutory right to adjudication 'at any time'; the Parties should be aware that the Housing Grants, Construction and Regeneration Act 1996, as amended, or the Construction Contracts (Northern Ireland) Order 1997, as amended,[ii] are likely to imply a right to adjudication via the appropriate Scheme in the absence of any express provisions. The leap of faith needed to form a true alliance contract using any standard form will always be a nervous one for those used to codified dispute procedures. We anticipate seeing some Option Z amendments creeping into alliance contracts, and our work with leading legal advisers suggests that such developments are being actively contemplated.

A dispute escalation ladder is therefore generally available in most NEC4 contracts, though the term 'ladder' is not used. The expectation is that you start small, quick and cheap in terms of resolving disputes. If that does not work, you move on to the most time- and money-absorbing processes, like litigation, although we must highlight the rising costs and resources needed to participate in adjudication. Subject to local laws, the Parties are free, however, to remove, change or add dispute resolution procedures by agreement. While the drafters chose those procedures carefully to suit each form of NEC4 contract, you can never cater for all eventualities; sometimes it may well be prudent to use the likes of mediation or conciliation. It would be a shame to bind yourself to something that does not quite fit the dispute in question, so always, even when the Parties are in a significant dispute, the Parties should just sensibly agree that the contract procedure is likely to work in that instance. If this is doubted, then change it. In most NEC4 contracts, there is a clause permitting the Parties to change the contract,[iii] though of course it would be prudent to seek legal advice before doing such a thing.

82. What procedures are used in what NEC4 contracts?

Table 8.1 is an illustration of the different types of dispute procedure, arranged, in our opinion, from the generally quickest or cheapest on the left-hand side through to the most expensive and lengthy procedures on the right. In some of the contracts, the drafters instruct one to try this procedure then move on to another; in some there are no procedures, as we mentioned. In the ALC, for instance, there is a simple choice of two procedures, with nowhere to go should the dispute not be resolved.

A few points on Table 8.1. With the ALC, strictly, it falls to the Alliance Board to resolve a dispute, but the board can choose to refer the dispute to an independent expert for an opinion or to the *Senior Representatives* of each member of the Alliance in dispute to help them resolve it. The *Senior Representatives*, in turn, may jointly appoint a mediator to assist them in the resolution of any issues. Both authors of this book are accredited mediators and understand the benefit of this underused (at least in the construction industry) method of resolving disputes. Contractual clauses can only really suggest a move to mediation as, at its heart, it is a consensual process. We have both seen the effects of a party attending an adjudication only because it was obliged to, as opposed to wanting to. This is a waste of time and money and should be avoided. The role of the independent expert in the ALC is like the requirements of the UK legislation, but does not comply fully with it; hence, our comments earlier about the statutory implied terms.

Across the NEC4 contracts, in Options W1 and W2, the Parties decide the *tribunal* when entering into a contract. This would generally be a choice between arbitration and litigation. Where Option W2 is incorporated, the use of *Senior Representatives* before adjudication is by agreement of the Parties, thus keeping the contract consistent with the requirements of the Housing Grants, Construction and Regeneration Act 1996, as amended, or the Construction Contracts (Northern Ireland) Order 1997, as amended, in the UK. Of course, if both Parties agree, earlier choices about dispute resolution may be amended.

83. Summary

The NEC4 contracts effectively have a dispute escalation ladder, though it is not called that. The drafters have created a list of dispute resolution procedures, these being a mix of the necessary (for jurisdictional reasons) and the desirable (to be in keeping with NEC principles). The drafters then consider this list against each of the NEC4 contracts in turn to attempt to be appropriate and consistent.

Table 8.1 Dispute procedures used in NEC4 contracts

NEC4 contract	None	Independent expert	Senior Representatives	Mediation	Adjudication	Dispute Avoidance Board	Arbitration	Litigation
Alliance Contract		x	x	x				
Design Build and Operate Contract, Option W1 or W2			x		x		x	x
Dispute Resolution Service Contract	x							
Engineering and Construction Contract, Option W1 or W2 or			x		x		x	x
Engineering and Construction Contract, Option W3						x		
Engineering and Construction Subcontract, Option W1 or W2			x		x		x	x
Engineering and Construction Short Contract					x		x	x
Engineering and Construction Short Subcontract					x		x	x
Framework Contract	x							
Professional Service Contract			x		x		x	x
Professional Service Short Contract			x		x		x	x
Professional Service Subcontract					x		x	x
Supply Contract			x		x		x	x
Supply Short Contract					x		x	x
Term Service Contract, Option W1 or W2			x		x		x	x
Term Service Subcontract, Option W1 or W2			x		x		x	x
Term Service Short Contract					x		x	x

Users are free to add their own dispute resolution clauses as Z clauses if they so desire, and they often do. It may be that they just wish to add mediation to an Engineering and Construction Short Contract (ECSC) or an independent expert to ECC Option W3, so it is quite an easy task to reproduce similar clauses, taking care as always with contract amendments.

Users are reminded that whatever NEC4 contract is used and, therefore, whatever dispute resolution procedures the Parties have, these are formal and agreed clauses. There really is nothing to stop the Parties seeking to resolve such issues as quickly and cheaply as possible. Advisers may very well say 'You should exert this right…' or 'You will not lose this case…' and, as far as legal advice is concerned, they may be correct. Ethics demands that professional advice is given in the interest of the client, but it is the Parties who have the dispute and, while they will listen to advice, they are free to direct their approach to a dispute as they see fit.

NOTES
i *Costain Limited* v *Tarmac Holdings Limited* [2017] EWHC 319 (TCC).
ii HMG (Her Majesty's Government) (1996) Housing Grants, Construction and Regeneration Act 1996. The Stationery Office, London, UK. Amended by HMG (2009) Local Democracy, Economic Development and Construction Act 2009. The Stationery Office, London, UK.
iii See, for example, Clause 12.3 of the ECC.

Chapter 9
The *tribunal* – arbitration or litigation?

84. Resolving disputes through arbitration or litigation

Unless the contract is being negotiated, which is probably a small percentage of NEC contracts awarded to date, it will generally be the *Client* that determines the *tribunal* and specifies it in the Contract Data. In many countries, both the courts and arbitration are available to decide disputes. Both have existed for resolving disputes for many years. According to Smith,[i]

> Archaeologists have uncovered evidence of the use of arbitration in the ancient civilizations of Egypt, Mesopotamia, and Assyria. Arbitration was extensively used by the ancient Greeks and Romans and in a form substantially similar to that used today.

As far as courts are concerned,[ii]

> The common law of England was largely created in the period after the Norman Conquest of 1066.

The end of the road of disputes is likely to be arbitration or the courts; there is nowhere to go after that. But how efficient are these processes and will justice be served and the 'right' answer always be reached?

A mix of opinions from a basic internet search throws these comments up, criticising arbitration and the courts.

- 'The law does not guarantee justice.'[iii]
- 'The president of the New South Wales Court of Appeal has called lawyers out on making "nit-picking arguments" that tie up courts and add to clients' bills.'[iv]
- 'I do find it unfortunate that taxpayer money has continued to be frivolously spent on this issue even after previous rulings from the Court of Appeals and the state Supreme Court.'[v]
- 'We have to have a certain tolerance for imperfection in our judges' (Alice Woolley, University of Calgary).[vi]
- 'The Wisconsin Lawyers' Fund for Client Protection – a little-known body created 30 years ago by the state Supreme Court – issued payments after lawyers pocketed unearned fees and embezzled client funds, among other acts.'[vii]
- 'Arbitration is increasingly under pressure, criticised for being time-consuming, costly and excessively formalised.'[viii]
- 'Arbitration: good decision or bad, you get what you bargained for.'[ix]

However, another mix, also from a basic internet search, throws up some counter opinions.

- 'When I started in practice [1959], it was an almost universal article of faith that English law and legal institutions were without peer in the world, with very little to be usefully learned from others' (Lord Bingham).[x]
- 'Brilliant Lord Bingham was the greatest judge of my time.'[xi]
- '[…] litigation is often a very long process and will not be officially ended for many years. Arbitration can be a much more efficient process. This is partly because some of the decisions that arbitrators make are not able to be appealed.'[xii]
- 'Arbitration can offer several advantages as an alternative to litigation:
  - Flexibility – The form and type of arbitration can be tailored to suit the parties.
  - Speed – The process can be started and resolved quickly, without waiting for court dates. Discoveries and preliminary processes are kept to a minimum.
  - Efficient – Although the parties must pay the costs of the arbitration, it is often more efficient than litigation in the courts.
  - Confidential – With few exceptions, proceedings take place in private and awards are not published without the consent of the parties.
  - Voluntary – Arbitration takes place only by the parties' mutual consent. This consent may be given when the parties enter a contract, or later when the dispute arises.
  - Final – The arbitrator's decision is final and binding, and court appeals are rare.'[xiii]

A mixed set of advantages and disadvantages for arbitration and the courts. One thing that most commentators would agree on is the uncertain nature of both processes. So many things are variable, such as the competence of the advocates, compelling position statements, helpful and honest experts, reliable evidence from witnesses and so on. For parties to put their faith in either process, where so much can go wrong, might be a brave move. Better then not to lose control and use a different approach, where parties stay in control. But arbitration, despite its many advantages, allows an appeal to the courts on a point of law so many parties believe that a court decision is likely to be more certain than one from an arbitrator.

Most commentators would probably see arbitration as mainstream these days, certainly in cross-border trade, and not forming part of a growing number of alternative dispute resolution processes available to disputing parties.

The main benefit of having a dispute decided in a court of law must be that justice should be served. The right answer should be achieved. On apparently extremely similar facts of a few construction cases, courts in England[xiv] seem to have a different opinion from courts in Scotland,[xv] reflecting the two countries' different legal systems The court structures and appeal courts also mean that you might not always get the 'right' answer at the first court. There have been plenty of yes/no/yes etc. type decisions over the years, and even split decisions at the very highest level of courts in the UK. Look at a famous swimming pool case[xvi] for a yo-yo set of decisions from highly competent arbitrators and judges. So, your adviser may well be pushing you along, saying you have a great chance, the evidence is good, the witnesses are good and so on, but are things as clear cut as predicted? It does not take too much to completely derail a potentially winning case.

In England and Wales, there is a specialist court set to deal with disputes about buildings, engineering, IT and surveying.[xvii] You might be in the camp that says this is helpful, as the judges and the counsel are well versed in these types of case. You might be in the camp that says this is regrettable. Why not eliminate disputes or resolve them before they reach the courts, rather than have very busy courts dealing with problems from the construction industry?

The stress, time to decision, cost and uncertainty should all play a big part in anyone deciding to go down the 'day in court' route, whatever the advice of advisers.

The main benefit of having a dispute decided in arbitration must be that it is conducted in private. Parties are not embarrassed by their antics leading up to the dispute being heard by legal commentators or even the press. It may be that their contract is sensitive or that washing dirty linen in public may be perceived as being off-putting to future partners, so arbitration is often taken to be the preferred route here. Not that it should be a worse option; arbitrators are usually exceptionally well-trained in the law of evidence and managing arbitration cases. They will often also be technically qualified so can understand the issues in dispute more readily than many judges.

Both arbitration and the courts appear though to be expensive and time-consuming, compared with other alternative dispute resolution processes, but we suggest they are the processes most likely to lead to determination of the facts in any dispute and that arbitrators and judges will be able to apply the law accordingly. This would be expected with the painstaking approach to proving evidence that each take.

Maybe, these days, neither the courts nor arbitration are seen to be dynamic enough for most modern businesses. Everyone is in a rush, even to get justice, and waiting months, possibly many months, to get your money is just not acceptable. Businesses might possibly prefer less money but more quickly, with a bit of rough justice, than more money in a wrong financial year. Maybe this is a nice research idea for someone?

Arcadis[xviii] found in 2017 that the global average value of disputes was US $43.4 million, where they define a 'dispute' as a situation where two parties typically differ in the assertion of a contractual right, resulting in a decision being given under the contract, which in turn becomes a formal dispute. They refer to the value of a dispute as being the entitlement, additional to that included in the contract, for the additional work or event which is being claimed. Arcadis also found that the global average length of disputes was nearly 15 months, where the length of a dispute is the period between it becoming formalised under the contract and the time of settlement, or the conclusion, of the hearing.

Arcadis also found that the number one cause of construction disputes was the failure to properly administer the contract.[xix]

In determining the most popular alternative dispute resolution methods to resolve disputes, Arcadis[xx] found that party-to-party negotiation was first, followed by mediation then arbitration. These were global findings; in the UK, Arcadis found that the order was adjudication (contract or ad hoc), party-to-party negotiation then mediation.[xxi] This, of course, most probably reflects the statutory basis of adjudication found in the UK.

In an article in the *New Statesman*,[xxii] it was noted that austerity cuts by the UK Ministry of Justice have led to the closure of 230 crown, county and magistrates' courts since 2010.

A convenient definition of mediation is as follows: a process conducted by an independent third party, in a strictly confidential manner, where the objective is to facilitate the parties resolving their dispute.[xxiii]

In order to understand properly the mediation process, it is essential that those involved in it fully appreciate that the role of the mediator is not to broker a settlement between the parties, but rather to assist those parties in negotiating their own settlement of the dispute. One important difference between mediation and more formal dispute resolution processes, such as litigation, arbitration or adjudication, is that the parties retain control over the dispute and its settlement. Another essential feature of the mediation process is its confidentiality.

Users of construction contracts should bear in mind that routine, or lower-value, cases are unlikely to be heard in specialist courts; therefore, you will be reliant on some of these courts that have suffered funding reductions.

Costs, speed and privacy are but a few reasons for favouring adjudication and mediation over litigation. Both authors of this book are accredited mediators and would be quite happy with increased use of mediation! Perhaps the courts should be viewed as the absolute last resort, when you are on your knees with nowhere else to go.

85. Avoiding disputes through arbitration or litigation

So, to avoiding disputes through arbitration or litigation. Once the 'go' button is pressed in arbitration or litigation, parties tend to take a supporting role and their advisers take over. Basically, the parties have lost some element of control at this point and everything must be done according to the rules, which are enforced by the arbitrator or judge. All opportunities to have one last face-to-face negotiation, as tough as it might be, have gone. That said, 'settlement on the steps of the court' is still a popular saying and still has a place in deciding disputes. Albeit, this is most likely to be incredibly rushed and off the cuff, compared with a little bit more of a relaxed atmosphere in the likes of mediation.

The fear of going to arbitration or litigation, after all the threats and the posturing, is probably the most likely reason that the event will be avoided. While judges, arbitrators and parties' advisers might be at pains to make sure the process is followed, and every bit of evidence carefully considered, parties quickly realise they have lost control of the dispute and perhaps anything can happen. The whole dispute may turn on a horrendous display of evidence-giving from a key witness, or a so-called expert blatantly and wrongly offering up partisan evidence on behalf of the party who has appointed them. The parties are absolutely miles away from their comfort zones and the days on end spent listening to everything being said is frustrating, stressful and nothing to do with their core businesses. That is when parties probably think 'What have I done?'

We have not much to say about the courts or arbitration in this book, certainly in terms of trying to avoid disputes. Once parties are hooked in to either of these processes, all you can do is trust in your advisers and hope that your evidence and arguments are good enough to impress the necessary people. In a recent English case, the judge wanted to highlight it as being crazy, as society pays. As we are told that the judge comes free in a court of law, one of the perceived benefits of the courts, just who paid for that decision then? Society did, of course; it came from taxpayers. In England and Wales, revised court fees have rebalanced that allocation and parties face a greater share of the burden.

Those advisors encouraging a party to proceed with arbitration or litigation are hopefully, giving the very best advice they can, before you commit to starting proceedings.

NOTES

i Smith RM (1998) *ADR for Financial Institutions*, 2nd edn. West Group, Eagon, MN, USA.
ii Kiralfy AR, Clendon MA and Lewis ADE (2019) Common Law. https://www.britannica.com/topic/common-law (accessed 21/01/2019).
iii Berkun S (2013) The Law Does Not Guarantee Justice. http://scottberkun.com/2013/the-law-does-not-guarantee-justice/ (accessed 21/01/2019).
iv Mezrani L (2015) Judge Warns Lawyers Not to Waste Court Time and Client Money. https://www.lawyersweekly.com.au/news/16470-judge-warns-lawyers-not-to-waste-court-time-and-client-money (accessed 21/01/2019).
v Gerould SA (2014) Did Court Cases Waste Taxpayer Money? Councilman Involved Says 'Yes'. http://www.uticaod.com/article/20140915/news/140919676 (accessed 21/01/2019).
vi CBC Radio (2017) There's a Difference Between Bad Judges and Bad Decisions. https://www.cbc.ca/radio/the180/stop-subsidizing-seniors-good-judges-can-make-bad-decisions-and-which-canadian-city-is-the-most-american-1.4028473/there-s-a-difference-between-bad-judges-and-bad-decisions-1.4028575 (accessed 21/01/2019).
vii Spivak C (2011) $2 Million Paid After Lawyer Misconduct. http://archive.jsonline.com/watchdog/watchdogreports/2-million-paid-after-lawyer-misconduct-4g3fg4k-136046248.html/ (accessed 21/01/2019).
viii Department of Private Law, University of Oslo (2013) The Future of Arbitration: Still Efficient and Preferred Choice? https://www.jus.uio.no/ifp/english/research/projects/choice-of-law/events/2013/130527-seminar.html (accessed 21/01/2019).
ix Richman RL (2010) Arbitration: Good Decision or Bad, You Get What You Bargained For. http://www.bullivant.com/Arbitration-eAlert (accessed 21/01/2019).
x Bingham T (2000) 'There is a world elsewhere': the changing perspectives of English law. In *The Business of Judging: Selected Essays and Speeches*. Oxford University Press, Oxford, UK, pp. 87–102.
xi Hirsch A (2010) Brilliant Lord Bingham Was the Greatest Judge of My Time. https://www.theguardian.com/law/afua-hirsch-law-blog/2010/sep/12/lord-bingham-civil-liberties (accessed 21/01/2019).
xii Hutchison & Stoy (2017) What is Arbitration? https://www.warriorsforjustice.com/what-is-arbitration/ (accessed 21/01/2019).
xiii ADR Institute of British Columbia (2018) Why Arbitration. https://adrbc.com/BCAMI/Arbitration/Benefits_of_Arbitration.aspx (accessed 21/01/2019).
xiv *Henry Boot Construction (UK) Ltd* v *Malmaison Hotel (Manchester) Ltd* (1999) 70 Con LR 32.
xv *City Inn Limited* v *Shepherd Construction Limited* [2010] ScotCS CSIH 68.
xvi *Ruxley Electronics and Construction Ltd* v *Forsyth* [1995] UKHL 8, [1996] AC 344.
xvii Her Majesty's Courts & Tribunals Service (2019) Technology and Construction Court. https://www.gov.uk/courts-tribunals/technology-and-construction-court (accessed 21/01/2019).
xviii Arcadis (2018) *Global Construction Disputes Report 2018: Does the Construction Industry Learn From Its Mistakes?* Arcadis, London, UK, p. 8.
xix Arcadis (2018) *Global Construction Disputes Report 2018: Does the Construction Industry Learn From Its Mistakes?* Arcadis, London, UK, p. 10.
xx Arcadis (2018) *Global Construction Disputes Report 2018: Does the Construction Industry Learn From Its Mistakes?* Arcadis, London, UK, p. 11.
xxi Arcadis (2018) *Global Construction Disputes Report 2018: Does the Construction Industry Learn From Its Mistakes?* Arcadis, London, UK, p. 17.

xxii Maguire P (2018) Crumbling Britain: The quiet decline of English courts. *New Statesman*, 27 Jun. See https://www.newstatesman.com/politics/uk/2018/06/crumbling-britain-quiet-decline-english-courts (accessed 21/01/2019).

xxiii Evans R (2016) Mediation. In *Keating Construction Dispute Resolution Handbook, Third edition*. ICE Publishing, London, UK, pp. 99–116.

Gerrard and Waterhouse
ISBN 978-0-7277-6404-1
https://doi.org/10.1680/necrad.64041.117
ICE Publishing: All rights reserved

Chapter 10
The *Senior Representatives*

Introduced for the first time to some of the NEC4 contracts, this can probably be more of a dispute avoidance process than one of dispute resolution. Although the aim of using *Senior Representatives* is to resolve any dispute put to them, it is not stated that they act as dispute resolvers. It equally does not prevent them from acting in this role either. They are completely free to attempt to resolve the dispute or get the dispute resolved by others, within the time allowed.

In the ECC, whether using Option W1 or W2, the use of *Senior Representatives* follows a similar path. In Option W1, the Parties are first obliged to use this process; if the matter is not resolved by the *Senior Representatives*, it is referred to and decided by the *Adjudicator*.[i]

There are a few differences in the use of this process between Options W1 and W2. The first is that in Option W2 the Parties need to agree that a dispute arising under or in connection with the contract is referred to the *Senior Representatives*.[ii] This is done to preserve the Parties' right under UK statutes to adjudication for certain contracts caught by the Housing Grants, Construction and Regeneration Act 1996, as amended, and its equivalent in Northern Ireland.[iii] This drafting style is followed in most other main NEC4 contracts using Option W2. In Option W2, a Party may refer a dispute to the *Adjudicator* at any time, even if the dispute has been referred to the *Senior Representatives*.[iv] In Option W1, a dispute arising under or in connection with the contract is referred to the *Senior Representatives* in accordance with the Dispute Reference Table. No such table exists in Option W2; basically, any dispute at any time can be referred to the *Adjudicator*, as mentioned before. The table in Clause W1 has time limitations for referring disputes to the *Senior Representatives*. For example, if the *Project Manager* has taken an action, such as to conduct a *Project Manager's* assessment of a compensation event, with which the *Contractor* is not happy, that dispute must be referred within four weeks of the *Contractor* becoming aware of the action. The *Senior Representatives* will need to be satisfied in the first instance that they can act as provided for in the contract when Option W1 is used.

The NEC4 ALC provides for the *Senior Representatives* differently, in that the Alliance Board has a choice. Clause 95.2 allows for the Alliance Board to choose between referring a dispute to an independent expert for an opinion or to the *Senior Representatives* of each member of the Alliance in dispute. The Alliance Board is free to ignore all or part of what comes back from the independent expert or *Senior Representatives* as appropriate. Nowhere is it stated that the Alliance Board is bound by either process. It may be that the *Senior Representatives* can quite quickly agree that their dispute is of such a nature that perhaps only arbitration or litigation can give it the justice it deserves. Or maybe it is of such inconsequential amounts that they simply negotiate. That is the beauty of this process; between the Parties and the *Senior Representatives*, they remain in control of their dispute and take it in whichever direction they feel is appropriate, unlike most other forms of dispute resolution, where parties lose control and are in the hands of others, making sure the requirements of the process are very carefully followed.

So just what do the *Senior Representatives* do then? Who are they, are they supposed to be neutral and where do they come from? In the ECC, the *Senior Representatives* are stated in Contract Data part one by the *Client*, at tender stage, and Contract Data part two by the *Contractor*, again at tender stage. No minimum number of *Senior Representatives* to be provided, but we consider it naturally follows that there should be the same number from each of the *Client* and the *Contractor*. Both parts of the Contract Data contain provision for two *Senior Representatives*, but, as with other parts of the Contract Data, this is entirely flexible and there could be more or less. For most biparty contracts, probably two each from the *Client* and the *Contractor* would be sensible. Perhaps one from each might be a problem, with personality clashes, diary issues etc., so four in total might work better. There is no evidence on this as far as we are aware; this could be an interesting research project for a student in the future. Interestingly, there is no clause to deal with retirement, replacement and the like of *Senior Representatives*, so the Parties would have to reply on other provisions of the contract for this.[v]

As for neutrality, again the ECC is silent in this regard, but we consider the best outcomes might be achieved where the *Senior Representatives* have little or no input to the contract. If they do, they may very well have some vested interest, so getting them to act in good faith might be difficult. The *Senior Representatives* are not bound by Clauses 10.1 or 10.2; they are not mentioned in these clauses, but they will surely have to agree some sensible ground rules for their conduct, or else they may struggle to use the time wisely. We think they will almost certainly agree to act in good faith, a key requirement for the likes of mediation to be effective. But it is simply up to them; there are no 'rules' in the contract covering this. Acting in good faith, with no axe to grind, and having little staked personally on the outcome should allow the *Senior Representatives* to be effective. As for neutrality from the Parties themselves, we do not think it is wise to do this and the *Senior Representatives* should ideally come from the Parties. We want the *Senior Representatives* to care about the dispute and arrive at something workable, which no doubt can be internally justified. If this is handed over to people outside of the Parties' organisations, this would be quite a dangerous move, at least financially. As mediators, we have seen the positive impact of a fresh pair of eyes dropped in to the dispute to help achieve a settlement.

When choosing the likely candidates to act as *Senior Representatives*, the *Client* and the *Contractor* will each have to think about an appropriate skill set. They will ideally be senior, as indicated in the title. Somebody who is wise, a good listener, a good negotiator, trustworthy and realistic, confident, polite, technically proficient and honest would, we think, give the process the best chance of success. Someone who happens to be a good dispute resolver might not look at the dispute from the widest angle possible to arrive at something that both Parties can live with and move on. For that is the aim of mediation and we think closely matches what the *Senior Representatives* are about. If they sit and throw alleged facts, opinions and position statements at one another, the opportunity to resolve will quickly pass them by and the Parties in an ECC contract may as well have gone straight to adjudication, where the *Adjudicator* will sit and properly listen to such things.

What they do is to follow the provisions in the contract, found in Clause W1.1 or W2.1, as appropriate. The 'threat' to the *Senior Representatives* is time and outcome. If they cannot agree an outcome within the time allowed, the dispute may be referred to and decided by the *Adjudicator*. This should be uppermost in their minds; neither Party probably wishes to proceed to a more formal dispute resolution process, which probably has significant time, cost, cash flow and resource demands and no end of other implications, so can they get somewhere they can both live with? When a dispute occurs, generally one Party will decide to formalise it. That Party notifies the *Senior Representatives*, along with the other Party and the *Project Manager*, of the nature of the dispute it wishes to resolve. From that notification, the referring Party has just one week to submit its statement of case. The Parties can agree on the length of the statement of case, but if they do not the default is ten sides of A4 paper. No mention is made of font size, but hopefully the Parties are reasonable and provide up to ten single sides of A4 paper with a readable size of font! There can be supporting evidence; no limit is put on this. We do not think though that the *Senior Representatives* will thank their colleagues for burying them in a mass of information in an attempt to be persuasive!

Under Clause W1.1(3) the *Senior Representatives* then have quite a free hand on what they do next. They can hold one or many meetings. They can negotiate. They can decide to push the matter over to mediation or conciliation, or over to any other procedure, to try to resolve the dispute over a period of no more than three weeks. Hopefully, the first meeting is one of honesty, realism and pragmatism. They can reflect on the statement of case, at which point they only have that from the referring Party. They might set a week aside to obtain a similar document from the other Party. They might just agree perhaps that it is so complex and unusual that the best place for this is litigation, but that is not what the contract wants as an output from the *Senior Representatives*. They must 'try to resolve the dispute';[vi] that is their job. The *Senior Representatives* must have a go, at least an attempt, at negotiating.

When the time allowed is up, the *Senior Representatives* are obliged to produce a list of the issues agreed[vii] and these are put into effect by the *Project Manager* and the *Contractor*. If what is agreed is that the *Project Manager* should give an instruction to change the Scope, then that is what should happen. If it is agreed that the *Contractor* is to submit some design for acceptance that it considered it did not have to do, the *Contractor* puts that into effect. If the Parties need to change a term of the contract,[viii] the *Contractor* will have to initiate this and agree matters with the *Client*; the *Project Manager* does not have any authority to do this. At the end of the process, the *Senior Representatives* should also produce a list of the issues not agreed[ix] and either Party may initiate adjudication if it so chooses to.

For this entire process, Clause W1.1(4) provides that no evidence of the statement of case or any discussions during the three weeks can be used or referred to, should there be subsequent proceedings before the *Adjudicator* or the *tribunal*. This is a similar position to that when using mediation; the process is carried out in confidence and the *Senior Representatives* cannot later be held to something they said or did, perhaps in an off-guard moment.

It can be seen that the provisions of the contract are extremely brief in respect of what the *Senior Representatives* do. We think that is a good idea as to constrain them might mean missing out on allowing the *Senior Representatives* to have a real go at resolving a dispute in a timely and cost-effective fashion, avoiding all sorts of future problems and hardship.

Thinking about this chapter and the title of the book, the *Senior Representatives* have the job of 'creatively' trying to resolve any dispute within three weeks of receiving a statement of case and supporting evidence from the referring Party. We say 'creatively', as there is such a limited time allowed, which we support, and therefore the traditional dispute resolution processes just cannot be accommodated. The *Senior Representatives* have a completely open brief to resolve the dispute. The contract does not even use words like 'negotiate' or 'mediate', but they are some of the processes that are designed to bring time benefits to disputing parties. Obviously, the Parties should try to resolve their disputes even before the *Senior Representatives* act and it should be noted that this is entirely possible. The *Contractor* can turn to the *Client* and say 'We have a problem here, how would you like to sit down and discuss it with me in a bid to resolve it?' We have looked at the time allowed in Option W1 to refer a dispute to the

Senior Representatives, and while we note that, for some disputes, there could be a four-week period before a Party is compelled to refer the dispute, it does give the Parties some time to try a reach some sort of settlement within the time frame. The 'threat' therefore of even *Senior Representatives* getting involved, let alone the more formal dispute resolution processes, might be the thing that helps the Parties to avoid disputes in the first place.

NOTES
  i ECC, Clause W1.1(1).
  ii ECC, Clause W2.1(1).
  iii HMG (Her Majesty's Government) (1996) Housing Grants, Construction and Regeneration Act 1996. The Stationery Office, London, UK. Amended by HMG (2009) Local Democracy, Economic Development and Construction Act 2009. The Stationery Office, London, UK.
  iv ECC, Clause W2.2(1).
  v ECC, Clause 12.3.
  vi ECC, Clause W1.1(3).
  vii ECC, Clause W1.1(3).
  viii ECC, Clause 12.3.
  ix ECC, Clause W1.1(3).

Chapter 11
Adjudication in Option W1

We have written Chapters 11 and 12 as stand-alone chapters. There is, therefore, deliberate repetition; if a user only uses Option W1 or W2, it would be quite tricky to write about Option W1 and then describe the differences in Option W2. In fact, this style of writing is exactly how the drafters present us with Options W1 and W2; they are stand-alone, but are quite similar indeed.

We have main Options and secondary Options, but note that Clauses W1, W2 and W3 on resolving and avoiding disputes are just called Options. Option W1 is the method of dispute resolution when the Housing Grants, Construction and Regeneration Act 1996, as amended, or its Northern Ireland equivalent,[i] does not apply. Option W3 may also be used in these circumstances.

86. Avoiding disputes in Option W1 adjudication

It is fair to say that in the clauses that deal with adjudication in Option W1 there are no obvious provisions for avoiding disputes. That said, even if the Parties are in adjudication, they can still control their own fortunes in a dispute by reaching an acceptable settlement and jointly terminating the appointment of the *Adjudicator*.[ii] The adjudication clauses in Option W1 simply focus on what the Parties and the *Adjudicator* do, and by when. This is what we will focus on.

87. Resolving disputes in Option W1 adjudication

In Option W1, the Parties are obliged to first refer a dispute arising under or in connection with the contract to the *Senior Representatives*. Clause W1.1(1) states that if a dispute is not resolved by the *Senior Representatives*, it is referred to and decided by the *Adjudicator*. If the *Senior Representatives* do not resolve the dispute, it goes through the adjudication process, assuming that it has been referred by one of the Parties. Clause W1.1(1) also states that a Party may replace a *Senior Representative* after notifying the other Party of the name of the replacement'. This could deal with long short- or long-term illness, change of employment or simply to improve things by changing a personality in this group. We describe the *Senior Representatives'* role in Chapter 10.

88. The *Adjudicator*

Some argue that it makes sense for the *Adjudicator* to be identified in the contract at the outset. This is achieved simply by stating the name of the *Adjudicator* in the Contract Data. The Parties know who that person is; they do not have to wait for an *Adjudicator* to be appointed, so this process lends itself to being quick. The downside is that the Parties may have an *Adjudicator* who is a lawyer and a dispute about complicated ground conditions. Or the *Adjudicator* might be an engineer while the dispute is centred on a complicated matter of law. Adjudication is a time-limited process and the named person might not have the availability to decide on a dispute when asked, owing to holidays or pressure of work. Some people have also questioned whether someone named by the *Client* can genuinely be viewed without concern for unintended bias.

We prefer that the Parties obviously try to avoid the dispute as best they can, but in the event that they cannot, and that the *Senior Representatives* fail to resolve the dispute, they then select a competent adjudicator whose expertise matches the dispute in question. Where they cannot, the *Adjudicator nominating body* must make a nomination and the Parties must use that person. Within both Options W1 and W2, once an adjudicator is appointed, that person becomes the *Adjudicator* (Clause W1.2(3)) and is therefore in place should a second and subsequent dispute(s) arise.

The Parties are obliged to use the DRSC current at the *starting date* to appoint the *Adjudicator*. This allows for the possibility that the DRSC changes over time; the latest version at the *starting date* of the project should be used.

Note that the *Adjudicator* is not included in the Clause 10.1 and 10.2 requirements. Instead, Clause W1.2(2) requires the *Adjudicator* to act impartially. The same clause also requires the *Adjudicator* to decide on the dispute as an independent adjudicator and not as an arbitrator, reflecting the position in certain legal jurisdictions.

There is a process for the Parties to choose an *Adjudicator* where the Contract Data entry is not completed, or the current *Adjudicator* resigns or is unable to act.[iii] This clause goes on to involve the *Adjudicator nominating body*, which will nominate the *Adjudicator* for the Parties where the Parties fail to agree an appointment.

Quite a few *Adjudicator nominating bodies* are available for inclusion here, such as the Chartered Institute of Arbitrators, the Institution of Civil Engineers and the Royal Institution of Chartered Surveyors, to name but a few, and the Parties should inform themselves as best they can when deciding which body to include in the tender document. The *Client* will generally select this, but it may well be that the *Contractor* persuades the *Client* differently at tender stage. Whoever the *Adjudicator* is, the Parties have a right to expect a very able and competent person to act in this important role; this is generally the case, but there are occasional exceptions where the courts decline to enforce an *Adjudicator's* decision.

Saying that, do the Parties always get very able and competent expert witnesses, mediators, conciliators, arbitrators or judges? You would certainly like to think so, but the reality may be somewhat different.

If there is a replacement *Adjudicator*, Clause W1.2(4) confirms the powers that the replacement *Adjudicator* has, namely to deal with undecided disputes. Clause W1.2(5) states the limitation of available action against the *Adjudicator* and the *Adjudicator's* employees and agents, limiting this to bad faith; not incompetence, for example. So, select your intended *Adjudicator* very carefully indeed, or use a reputable *Adjudicator nominating body*.

89. The adjudication

The adjudication starts with a Party issuing a notice of adjudication to the other Party and the *Project Manager*. This follows shortly after the *Senior Representatives* have produced the list of agreed and not-agreed issues. The dispute is referred to the *Adjudicator* within one week of the notice of adjudication.[iv] The notice sets the jurisdiction of the *Adjudicator* and must therefore be drafted carefully. Where there is a discrepancy between the redress sought in the notice and that sought in later submissions, the notice takes precedence.

Clause W1.3(8) states that the *Adjudicator* decides on the dispute. When this is done, the *Adjudicator* informs the Parties and the *Project Manager* of the decision and reasons within four weeks of the end of the period for receiving information. The four weeks may be extended if the Parties agree.

So how does the *Adjudicator* decide on the dispute? The *Adjudicator* considers the evidence submitted by the Parties and reaches the decision, having analysed the facts and the law appropriate to the dispute referred. Although there is no standard format, the decision itself will usually be written up in a style fairly like what the Parties might receive from an arbitrator or a judge and must include reasons.[v] The decision will take the Parties through the evidence and the submissions and give reasons as to what evidence and which arguments were preferred, and why. This will hopefully give the Parties a good idea of where they went right or wrong and perhaps give grounds or reasons for taking the dispute on to the *tribunal*.

The end of the period for receiving information is linked to the date that the Party referred the dispute to the *Adjudicator*.[vi] From when the referral is made, any more information from a Party to be considered by the *Adjudicator* is provided within four weeks of the referral. While the date for reaching the decision may not be extended without at least one Party's consent, the *Adjudicator* is free to direct the timing of submissions without the Parties' involvement. Where the Parties are co-operating, the timetable and decision date can sometimes be agreed in a telephone conference call. Where the Parties do not seem to be using mutual trust and co-operation, one of the Parties is unlikely to agree to such an extension unless it is in its interests, whether tactically or in respect of the time needed to respond. Many adjudications elongate, and the *Adjudicator* sets a bespoke timetable.

Threaded into this might be meetings to allow the inquisitorial side of adjudication to flourish. The adjudication could use documents only or a more traditional style like litigation with claim, defence and counterclaim, all supported by verbal evidence. The *Adjudicator* is, though, trying to balance natural justice with the quick approach that adjudication is supposed to take.

It is possible for a subcontract dispute to be referred to the *Adjudicator* so that the disputes can be decided together; this is provided in Clause W1.3(4).

What the *Adjudicator* may do in the adjudication is stated in Clause W1.3(5). Note that 'may' is a discretionary term and it is for the *Adjudicator* to decide which, if any, of the listed provisions will help in the adjudication. Certainly, the second list item is a standout tool for the *Adjudicator*, in that the *Adjudicator* can take the initiative here and ascertain the facts and the law relating to the dispute. We like this provision, the 'roll your sleeves up and get stuck into the facts and the law' approach, rather than 'sit and listen', as found in the courts. If the *Adjudicator* is not sure of something, dig deeper, keep going until you get the information needed, drawing inferences if information requested is not provided. The decision may be quite different from the lines of argument made by both Parties in their referrals to the *Adjudicator*, based on the *Adjudicator* doing a proper job, which is to decide on the dispute. The Parties may well have created the dispute, but equally well may have not sufficiently argued the facts and the law to enable the *Adjudicator* to decide. The third list item may be useful; the *Adjudicator* can instruct a Party to provide further information related to the dispute. The fourth item helps with this too; a Party can be instructed by the *Adjudicator* to take an action, such as to prepare a quotation for a compensation event on a certain basis that the *Adjudicator* considers to be correct.

We rather applaud this style of drafting and the powers given to the *Adjudicator* by the wording in Option W1. The *Adjudicator* is there to do an important job, that is, to decide a dispute, which could have serious consequences for one or both Parties. The competent *Adjudicator* takes complete control of proceedings and takes the initiative in ascertaining the facts and the law related to the dispute. The *Adjudicator* must consider arguments, usually in writing, but possibly orally too at meetings. What we want and need here is a proactive, confident and competent person, able to take control and reach a decision within quite pressing timescales. This links to the care that the Parties need to take when selecting their *Adjudicator*.

A few other clauses in Option W1 regarding adjudication include making sure that communications are made to both Parties at the same time; such requirements are often reinforced by adjudicators in their initial directions to the Parties; independence needs to be shown.[vii] Should the *Adjudicator's* decision include additional cost or delay, the *Adjudicator* should follow the rules for assessing compensation events.[viii] The four-week period during which the *Adjudicator* must make a decision may be extended, but this would need agreement by the Parties.[ix] The *Adjudicator* might find that there is an overwhelming amount of evidence to work through or might have unfortunately suffered a bout of illness for a few weeks, so the *Adjudicator* would communicate with both Parties on this. There is a careful balance to achieve here. The decision needs to be provided in the four weeks, but for exceptional circumstances, because maybe to one of the Parties any delayed decision could have a financial consequence. Clause W1.3(9) requires the Parties, the *Project Manager* and the *Supervisor* to proceed as if the matter disputed was not disputed unless and until the *Adjudicator* informs the Parties of the decision. For example, a compensation event notified by the *Contractor* might not have been accepted as being a compensation event by the *Project Manager*. This might cause a dispute to arise and the *Contractor* might refer the matter to adjudication, having had no further success with the *Senior Representatives*. The *Adjudicator* might decide that the action of the *Project Manager* was wrong; therefore, after receiving the *Adjudicator's* decision, the *Project Manager* will now need to instruct the *Contractor* to submit a quotation under Clause 61.2.

Clause W1.3(10) states that the *Adjudicator's* decision will be binding on the Parties; the Parties must do as the *Adjudicator* decides. This is unless and until the decision is revised by the *tribunal*. The clause goes on to state that the *Adjudicator's* decision is enforceable as a matter of contractual obligation between the Parties and not as an arbitral award, meaning, therefore, that further proceedings will be needed at the *tribunal* for breach of contract if one of the Parties does not do what the *Adjudicator* decides. This same clause deals with the finality of the *Adjudicator's* decision if neither Party has notified the other of its dissatisfaction within the time allowed, along with a statement of intent to refer the matter to the *tribunal*. A Party also cannot refer a dispute to the *Adjudicator* that is the same as one that has already been referred to the *Adjudicator*. Perhaps the Party did not like the decision first time around, some other evidence has become available or some new case law supports the Party's previous position statements; well, that's tough, and the dispute cannot be referred.[x] Finally, if the *Adjudicator's* decision contains any clerical mistake or ambiguity, the *Adjudicator* may correct this, but only within two weeks of giving the decision to the Parties.[xi] This rule is commonly known as the slip rule. This might cover arithmetical errors, missing words that are needed to make sense of the decision or a statement in the decision that is made twice but with different outcomes.

90. How have the courts responded to adjudication?

The courts in the UK have generally been very supportive of adjudication, enforcing most adjudication decisions put to them.

91. Summary

Once the Parties take a dispute to adjudication, they lose control of the process. This dispute is now firmly in the hands of the *Adjudicator*, not in the hands of the Parties. The *Adjudicator* is charged with deciding on the dispute, including taking the initiative in ascertaining the facts and the law related to the dispute. An *Adjudicator* who considers that not all the information has been provided in order to decide on the dispute can instruct a Party to provide further information related to the dispute. This is an inquisitorial process and the Parties need a strong, competent person to act in this capacity. They should choose that person very carefully indeed. It is all too late now to attempt to avoid the dispute, the horses are running, but the Parties always have the right to negotiate a settlement and cancel the adjudication, even just before the *Adjudicator's* decision is given. Once the decision is given, the Parties are bound by it, unless and until it is revised by the *tribunal*. But the clock on when this must happen starts ticking as soon as the decision is given.

We consider adjudication to be a most appropriate form of alternative dispute resolution for many, but not necessarily all, disputes. The whole process is in keeping with the NEC philosophy of real-time management, making decisions and so on. We note the games being played in some adjudications in the UK and hope that competent adjudicators quash these or walk away from immoral ambush situations. It takes a brave but ethical person to do just that to protect adjudication and deliver timely justice to the Parties.

NOTES

i   HMG (Her Majesty's Government) (1996) Housing Grants, Construction and Regeneration Act 1996. The Stationery Office, London, UK. Amended by HMG (2009) Local Democracy, Economic Development and Construction Act 2009. The Stationery Office, London, UK.
ii See DRSC Clause 5.1.
iii ECC, Clause W1.2(3).
iv ECC, Clause W1.3(1).
v   ECC, Clause W1.3(8).
vi ECC, Clause W1.3(3).
vii ECC, Clause W2.3(6).
viii ECC, Clause W1.3(7).
ix ECC, Clause W1.3(8).
x   ECC, Clause W1.3(10).
xi ECC, Clause W1.3(11).

Chapter 12
Adjudication in Option W2

We have written Chapters 11 and 12 as stand-alone chapters. There is, therefore, deliberate repetition; if a user only uses Option W1 or W2, it would be quite tricky to write about Option W1 and then describe the differences in Option W2. In fact, this style of writing is exactly how the drafters present us with Options W1 and W2; they are stand-alone but quite similar indeed.

We have main Options and secondary Options, but note that Clauses W1, W2 and W3 on resolving and avoiding disputes are just called Options. Option W2 is the method of dispute resolution when the Housing Grants, Construction and Regeneration Act 1996, as amended, or its Northern Ireland equivalent applies.[i] In Option W2, time periods stated in days exclude Christmas Day, Good Friday and bank holidays,[ii] consistent with the legislation.

92. Avoiding disputes in Option W2 adjudication

It is fair to say that in the clauses that deal with adjudication in Option W2, there are no obvious provisions for avoiding disputes. That said, even if the Parties are in adjudication, they can still control their own fortunes in a dispute by reaching an acceptable settlement and terminate the appointment of the *Adjudicator*.[iii] The adjudication clauses in Option W2 just focus on what the Parties and the *Adjudicator* do, and by when. This is what we will focus on.

93. Resolving disputes in Option W2 adjudication

In Option W2, Clause W2.2(1) states that a Party may refer a dispute to the *Adjudicator* at any time, whether or not the dispute has been referred to the *Senior Representatives*. 'At any time' is an interesting statement and is exactly in line with the Housing Grants, Construction and Regeneration Act 1996, as amended,[iv] and its equivalent in Northern Ireland. But what does this mean? If the dispute occurred five or ten years ago and was never resolved, can a Party still refer the dispute to the *Adjudicator*? The answer possibly lies in legislation for limitation, which varies between the UK jurisdictions.

An option in W2, though, is that the Parties may choose to first refer a dispute arising under or in connection with the contract to the *Senior Representatives*, but only if both Parties agree to this.[v] If one Party wishes to assert is right not to use the *Senior Representatives* and instead go straight to the *Adjudicator*, then it may do so.[vi]

If the Parties do agree to first try to get the dispute resolved by the *Senior Representatives*, Clause W2.1(1) states that if a dispute is not resolved by the *Senior Representatives*, it is referred to and decided by the *Adjudicator*. Therefore, if the *Senior Representatives* do not resolve the dispute, it goes through the adjudication process, assuming it has been referred by one of the Parties.

Clause W2.1(1) also states that a Party may replace a *Senior Representative* after notifying the other Party of the name of the replacement. This could deal with long short- or long-term illness, change of employment or simply to improve things by changing a personality in this group.

94. The *Adjudicator*

Some argue that it makes sense for the *Adjudicator* to be identified in the contract at the outset. This is achieved simply by stating the name of the *Adjudicator* in the Contract Data. The Parties know who that person is, and they do not have to wait for an *Adjudicator* to be appointed, so this process lends itself to being quick. The downside is that the Parties may have an *Adjudicator* who is a lawyer and a dispute about complicated ground conditions. Or the *Adjudicator* might be an engineer while the dispute is centred on a complicated matter of law. Adjudication is a time-limited process and the named person might not have the availability to decide on a dispute when asked, owing to holidays or pressure of work. Some people have also questioned whether someone named by the *Client* can genuinely be viewed without concern for apparent bias.

We prefer that the Parties obviously try to avoid the dispute as best they can, but in the event that they cannot, and that the *Senior Representatives* fail to resolve the dispute, they then select a competent adjudicator whose expertise matches the dispute in question. Where they cannot, the *Adjudicator nominating body* must make a nomination and the Parties must use that person. Within both Options W1 and W2, once an adjudicator is appointed, that person becomes the *Adjudicator* (Clause W2.2(5)) and is therefore in place should a second and subsequent dispute(s) arise.

The Parties are obliged to use the DRSC current at the *starting date* to appoint the *Adjudicator*. This allows for the possibility that the DRSC changes over time; the latest version at the *starting date* of the project should be used.

Note that the *Adjudicator* is not included in the Clause 10.1 and 10.2 requirements. Instead, Clause W2.2(4) requires the *Adjudicator* to act impartially. The same clause also requires the *Adjudicator* to decide on the dispute as an independent adjudicator and not as an arbitrator, reflecting the position in certain legal jurisdictions.

There is a process for the Parties to choose an *Adjudicator* where the Contract Data entry is not completed, or the current *Adjudicator* resigns or is unable to act.[vii] This clause goes on to involve the *Adjudicator nominating body*, which will nominate the *Adjudicator* for the Parties where the Parties fail to agree an appointment.

Quite a few *Adjudicator nominating bodies* are available for inclusion here, such as the Chartered Institute of Arbitrators, the Institution of Civil Engineers and the Royal Institution of Chartered Surveyors, to name but a few, and the Parties should inform themselves as best they can when deciding which body to include in the tender document. The *Client* will generally select this, but it may well be that the *Contractor* persuades the *Client* differently at tender stage. Whoever the *Adjudicator* is, the Parties have a right to expect a very able and competent person to act in this important role; this is generally the case, but there are occasional exceptions where the courts decline to enforce an *Adjudicator's* decision.

Saying that, do the Parties always get very able and competent expert witnesses, mediators, conciliators, arbitrators or judges? You would certainly like to think so, but the reality may be somewhat different.

If there is a replacement *Adjudicator*, Clause W2.2(6) confirms the powers that the replacement *Adjudicator* has, namely to deal with undecided disputes. Clause W2.2(8) states the limitation of available action against the *Adjudicator* and the *Adjudicator's* employees and agents, limiting this to bad faith; not incompetence, for example. So, select your intended *Adjudicator* very carefully indeed, or use a reputable *Adjudicator nominating body*.

95. The adjudication

The adjudication starts with a Party issuing a notice of adjudication to the other Party with a brief description of the dispute and the decision which it wishes the *Adjudicator* to make.[viii] The referring Party needs to send a copy of the notice of adjudication to the *Adjudicator* when it is issued. Within three days of the receipt of the notice of adjudication, the *Adjudicator* is obliged to inform the Parties of one of two things: either that the *Adjudicator* can decide on the dispute in accordance with the contract or that the *Adjudicator* is unable to decide the dispute and has resigned. This is detailed in Clause W2.3(1), which includes is provision for where the *Adjudicator* does not so inform the Parties within the three days. The notice sets the jurisdiction of the *Adjudicator* and must therefore be drafted carefully. Where there is a discrepancy between the redress sought in the notice and that sought in later submissions, the notice takes precedence.

Clause W2.3(8) states that the *Adjudicator* decides on the dispute. When this is done, the *Adjudicator* informs the Parties and the *Project Manager* of the decision and reasons within 28 days of the dispute being referred. This clause goes on to state how this period may be extended.

In the *Adjudicator's* decision, the *Adjudicator* may allocate the *Adjudicator's* fees and expenses between the Parties.[ix]

So how does the *Adjudicator* decide on the dispute? The *Adjudicator* considers the evidence submitted by the Parties and reaches the decision, having analysed the facts and the law appropriate to the dispute referred. Although there is no standard format, the decision itself will usually be written up in a style fairly like that used by an arbitrator or a judge and must include reasons.[x] The decision will take the Parties through the evidence and the submissions and give reasons as to what evidence and which arguments were preferred, and why. This will hopefully give the Parties a good idea of where they went right or wrong and perhaps give grounds or reasons for taking the dispute on to the *tribunal*.

Within seven days of a Party giving a notice of adjudication, it must do three things. First, the Party must refer the dispute to the *Adjudicator* or arrange for an appointment of one. Next, it must provide the *Adjudicator* with the information on which it relies, including any supporting documents. Finally, it must provide a copy of the information and supporting documents it has provided to the *Adjudicator* to the other Party. This is all as provided for in Clause W2.3(2), which goes on to state that any further information from a Party to be considered by the *Adjudicator* must be provided within 14 days of the referral. While the date for reaching the decision may not be extended without at least one Party's consent, the *Adjudicator* is free to direct the timing of submissions without the Parties' involvement. Where the Parties are co-operating, the timetable and decision date can sometimes be agreed in a telephone conference call. Where the Parties do not seem to be using mutual trust and co-operation, one of the Parties is unlikely to agree to such an extension unless it is in its interests, whether tactically or in respect of the time needed to respond. Many adjudications elongate, and the *Adjudicator* sets a bespoke timetable.

Threaded into this might be meetings to allow the inquisitorial side of adjudication to flourish. The adjudication could be documents only or a more traditional style like litigation with claim, defence and counterclaim, all supported by verbal evidence. The *Adjudicator* is, though, trying to balance natural justice with the quick approach that adjudication is supposed to take.

It is possible for a subcontract dispute to be referred to the *Adjudicator* so that the disputes can be decided together; this is as provided in Clause W2.3(3).

What the *Adjudicator* may do in the adjudication is stated in Clause W2.3(4). Note that 'may' is a discretionary term and it is for the *Adjudicator* to decide which, if any, of these listed provisions will help in the adjudication. Certainly, the second list item is a standout tool for the *Adjudicator*, in that the *Adjudicator* can take the initiative here and ascertain the facts and the law relating to the dispute, albeit this wording is required by the UK statute. We like this provision, the 'roll your sleeves up and get stuck into the facts and the law' approach, rather than 'sit and listen', as found in the courts. If the *Adjudicator* is not sure of something, dig deeper, keep going until you get the information needed, drawing inferences if information requested is not provided. The decision may be quite different from the lines of argument made by both Parties in their referrals to the *Adjudicator*, based on the *Adjudicator* doing a proper job, which is to decide on the dispute. The Parties may well have created the dispute, but equally well may have not sufficiently argued the facts and the law to enable the *Adjudicator* to decide. The third list item may be useful; the *Adjudicator* can instruct a Party to provide further information related to the dispute. The fourth item helps with this too; a Party can be instructed by the *Adjudicator* to take an action, such as prepare a quotation for a compensation event on a certain basis that the *Adjudicator* considers to be correct.

We rather applaud this style of drafting and the powers given to the *Adjudicator* by the wording in Option W2. The *Adjudicator* is there to do an important job, that is, to decide a dispute, which could have serious consequences for one or both Parties. The competent *Adjudicator* takes complete control of proceedings and takes the initiative in ascertaining the facts and the law related to the dispute. The *Adjudicator* must consider arguments, usually in writing, but possibly orally too at meetings. What we want and need here is a proactive, confident and competent person, able to take control and reach a decision within quite pressing timescales. This links to the care that the Parties need to take when selecting their adjudicator.

A few other clauses in Option W2 regarding adjudication include making sure that communications are made to both Parties at the same time; such requirements are often reinforced by adjudicators in their initial directions to the Parties; independence needs to be shown.[xi] Should the *Adjudicator's* decision include additional cost or delay, the *Adjudicator* should follow the rules for assessing compensation events.[xii] Also, Clause W2.3(7) states that if the *Adjudicator's* decision changes an amount notified as due, the date on which payment of the changed amount becomes due is seven days after the decision. The 28-day period during which the *Adjudicator* must make a decision may be extended, but this would need agreement by the Parties.[xiii] The *Adjudicator* might find that there is an overwhelming amount of evidence to work through or might have unfortunately suffered a bout of illness for a few weeks, so the *Adjudicator* would communicate this to both Parties. There is a careful balance to achieve here. The decision needs to be provided in the 28 days, but for exceptional circumstances, because maybe to one of the Parties any delayed decision could have a financial consequence. Clause W2.3(9) requires the Parties, the *Project Manager* and the *Supervisor* to proceed as if the matter disputed was not disputed unless and until the *Adjudicator* informs the Parties of the decision. For example, a compensation event notified by the *Contractor* might not have been accepted as being a compensation event by the *Project Manager*. This might cause a dispute to arise and the *Contractor* might refer the matter to adjudication, having had no further success with the *Senior Representatives*. The *Adjudicator* might decide that the action of the *Project Manager* was wrong; therefore, after receiving the *Adjudicator's* decision, the *Project Manager* will now need to instruct the *Contractor* to submit a quotation under Clause 61.2.

Clause W2.3(11) states that the *Adjudicator's* decision will be binding on the Parties; the Parties must do as the *Adjudicator* decides. This is unless and until the decision is revised by the *tribunal*. The clause goes on to state that the *Adjudicator's* decision is enforceable as a matter of contractual obligation between the Parties and not as an arbitral award, but the courts in the UK jurisdictions have shown their willingness to support most adjudication decisions. This same clause deals with the finality of the *Adjudicator's* decision if neither Party has notified the other of its dissatisfaction within the time allowed, along with a statement of intent to refer the matter to the *tribunal*. A Party also cannot refer a dispute to the *Adjudicator* that is the same as one that has already been referred to the *Adjudicator*. Perhaps the Party did not like the decision first time around, some other evidence has become available or some new case law supports the Party's previous position statements; well, that's tough, and the dispute cannot be referred.[xiv] Finally, there is provision for a slip rule, as required by UK statute; if the *Adjudicator's* decision contains any clerical mistake or ambiguity, the *Adjudicator* may correct this, but only within two weeks of giving the decision to the Parties.[xv] This might cover arithmetical errors, missing words that are needed to make sense of the decision or a statement in the decision that is made twice but with different outcomes.

96. How have the courts responded to adjudication?

The courts in the UK have generally been very supportive of adjudication, enforcing most adjudication decisions put to them.

97. Summary

Once the Parties take a dispute to adjudication, they lose control of the process. This dispute is now firmly in the hands of the *Adjudicator*, not in the hands of the Parties. The *Adjudicator* is charged with deciding on the dispute, including taking the initiative in ascertaining the facts and the law related to the dispute. An *Adjudicator* who considers that not all the information has been provided in order to decide on the dispute can instruct a Party to provide further information related to the dispute. This is an inquisitorial process and the Parties need a strong, competent person to act in this capacity. They should choose that person very carefully indeed. It is all too late now to attempt to avoid the dispute, the horses are running, but the Parties always have the right to negotiate a settlement and cancel the adjudication, even just before the *Adjudicator's* decision is given. Once the decision is given, the Parties are bound by it, unless and until it is revised by the *tribunal*. But the clock on when this must happen starts ticking as soon as the decision is given.

We consider adjudication to be a most appropriate form of alternative dispute resolution for many, but not necessarily all, disputes. The whole process is in keeping with the NEC philosophy of real-time management, making decisions and so on. We note the games being played in some adjudications in the UK and hope that competent adjudicators quash these or walk away from immoral ambush situations. It takes a brave but ethical person to do just that to protect adjudication and deliver timely justice to the Parties.

NOTES

i HMG (Her Majesty's Government) (1996) Housing Grants, Construction and Regeneration Act 1996. The Stationery Office, London, UK. Amended by HMG (2009) Local Democracy, Economic Development and Construction Act 2009. The Stationery Office, London, UK.

ii ECC, Clause W2.2(2).

iii See DRSC Clause 5.1, but this must be joint.

iv Section 108(2)(a) of the Housing Grants, Construction and Regeneration Act 1996, as amended.

v ECC, Clause W2.1(1).

vi ECC, Clause W2.2(1).

vii ECC, Clause W2.2(6).

viii ECC, Clause W2.3(1).

ix ECC, Clause W2.3(8).

x ECC, Clause W2.3(8).

xi ECC, Clause W2.3(6).

xii ECC, Clause W2.3(7).

xiii ECC, Clause W2.3(8).

xiv ECC, Clause W2.3(10).

xv ECC, Clause W2.3(11).

Chapter 13
The Dispute Avoidance Board in Option W3

In keeping with the dispute avoidance principles generally found within NEC contracts, it is probably no surprise to users that the drafters introduced a Dispute Avoidance Board in NEC4 contracts in 2017, as opposed to a Dispute Resolution Board. Perhaps the only surprise is that such a process did not feature in earlier versions of NEC, as it has been around for quite a while now in other forms of contract, most noticeably in FIDIC[i] and World Bank Procurement forms. Further, this provision is included only in the ECC,[ii] as the drafters consider that the Parties using this contract would be most likely to consider using such a process. Perhaps surprising is that the Dispute Avoidance Board is not offered in the ALC, with mediation being chosen instead where disputes arise. That said, the ALC has a 'no sue' style clause, so the usual NEC *tribunal* does not exist anyway. We believe that the ALC might benefit from the addition of Option W3.

Dispute boards in other forms of contract have generally relied on contractual status, as opposed to statute. Therefore, the enforcement of dispute boards' decisions has sometimes proved difficult. By limiting the role of the Dispute Avoidance Board to making recommendations, its role is somewhat restricted. Its recommendations don't need to be acted on, although it is probably preferable that they are. Consequently, the issue of enforcement doesn't arise. But the role of the Dispute Avoidance Board does have a hard edge to it; reference to the *tribunal* isn't possible until the board has made a recommendation.

The note at the start of Option W3 says that this Option should be used when 'a Dispute Avoidance Board is the method of dispute resolution and the United Kingdom Housing Grants, Construction and Regeneration Act 1996 does not apply'.[iii] The drafters, of course, realise that Option W3 does not comply with this piece of British legislation (or its Northern Ireland equivalent) and have commented on this accordingly. Interestingly, we note that several UK clients also realise this but, despite this, put their faith in the Dispute Avoidance Board process even though they know that one Party could revert to adjudication as provided for by the legislation and ignore the provisions of the contract.

The emphasis of the Dispute Avoidance Board is exactly as the title suggests; it is about avoiding and not resolving any disputes that arise. In fact, the Dispute Avoidance Board's primary role is found in Clause W3.2(1), which is to assist 'the Parties in resolving potential disputes before they become disputes'. This is further supported in Clause W3.2(5), where the Dispute Avoidance Board helps the Parties to settle all potential disputes 'without the need for the dispute to be formally referred'.

Though users might have expected the Dispute Avoidance Board to be an identified term, this would not be possible where the use of three Dispute Avoidance Board members is selected by the *Client*. When compiling the tender document, the *Client* selects whether the number of Dispute Avoidance Board members will be one or three and identifies this in the Contract Data part one. Where the number of Dispute Avoidance Board members is three, the third member is jointly chosen by the Parties. There is no provision in Option W3 for when the Parties must choose this third member by, but it can be easily implied that this must be before the first visit to the Site by the Dispute Avoidance Board takes place.

The Dispute Avoidance Board members are appointed by the Parties using the DRSC, as stated in Clause W3.1(2). Note that this does not say that it is the NEC4 version but instead the version current at the *starting date* (of the contract). This allows for the possibility of later versions of the DRSC to be published and therefore used. Parties should also remember that people aren't necessarily in position permanently. People retire, become ill, take sabbaticals, become pregnant, are busy etc. Therefore, the members of a board, particularly on longer-duration projects, may change over time.

It is possible, of course, that the Parties cannot agree on whom to appoint as the third Dispute Avoidance Board member, or similarly cannot agree on a replacement, where a member of the Dispute Avoidance Board is unable to act. Clause W3.1(4) allows for either Party to ask the *Dispute Avoidance Board nominating body* to choose one. The *Dispute Avoidance Board nominating body* is identified in Contract Data part one and is the sort of service that the Chartered Institute of Arbitrators,[iv] the Institution of Civil Engineers,[v] the Royal Institution of Chartered Surveyors[vi] and others can offer to users. A note of caution, though, is that these bodies seem to refer to dispute boards rather than Dispute Avoidance Boards; users would be wise to check that the 'avoidance' specialism can be provided. The *Dispute Avoidance Board nominating body* must choose a Dispute Avoidance Board member within seven days of a request from either Party; this person becomes a member of the Dispute Avoidance Board.

So why have one or three members of the Dispute Avoidance Board and not two? Certainly, the odd number may help the Dispute Avoidance Board members where a vote needs to be cast, perhaps as to what the recommendation for resolving a potential dispute might be, where the Parties have not resolved this by the end of the Site visit. Perhaps three members with complementary but different skills would be required for a large multidisciplinary project. Perhaps with only one member this is not the case, or there is a person has a most impressive range of skills to bring to the project. Or perhaps the matter is just one of pragmatism, consistency and cost. It is for the *Client* to decide and then for each tenderer to be satisfied with this when submitting a tender.

The skill set of the members needs careful consideration. Both technical and legal skills are required. But the softer skills of members are also important. For the recommendations of the board to be accepted by the Parties, the members of the board will need to be respected. Some may come with a professional reputation developed over time. Others may not, but will develop a relationship with the Parties during routine visits. The example demonstrates the value of these skills.

Example 13.1

This is a true story.

In a large infrastructure project involving the design and construction of several billion pounds worth of civil engineering works, dispute boards were deployed on most of the project contracts. For one contract, the board visited the site regularly and got to know the contractor's and the employer's people very well. The project came close to completion before the first potential dispute arose.

The board members were summoned to the site and heard a presentation in the site conference room before being taken to a remote corner of the site. They were shown a farmer's access track, poorly constructed and flooded. The problem included issues of design and construction, but probably only concerned £50,000 at most. The contract in question had a value of well over £100 million.

One of the board members, a well-known figure in construction dispute circles, expressed his annoyance at having been required to come and deal with such a trivial issue, in comparison with the wider achievements of the project. He demanded to be returned to the site offices, whereupon he left and returned home.

He called the contractor a day later and enquired if the matter had been resolved by the parties. He was told that it had been, and an apology was proffered to him for wasting his time.

This example is rare but shows the value of using people with the right soft skills. These must be considered, alongside the usual technical, legal and dispute resolution skills.

Having decided on the use of one or three members, the Dispute Avoidance Board needs to realise its role and how it goes about fulfilling it. We have identified what we consider to be the primary role already, but the Dispute Avoidance Board must do all of the following.

- Act impartially (Clause W3.1(3)).
- Visit the Site as required in Clause W3.1(5).
- Decide the agenda for the Site visit (Clause W3.1(6)).
- Assist the Parties in resolving potential disputes before they become disputes (Clause W3.2(1)).
- Visit the Site and inspect the *works* (Clause W3.2(5)).
- Review all potential disputes and help the Parties to settle them without the need for the dispute to be formally referred (Clause W3.2(5)).
- Prepare a note of the visit (Clause W3.2(5)).
- Provide a recommendation for resolving a potential dispute where the Parties have not resolved a potential dispute by the end of the Site visit (Clause W3.2(5)).
- Take the initiative in reviewing potential disputes, including asking the Parties to provide further information (Clause W3.2(6)).

The regular (and possibly ad hoc) visits to the Site provide the board with an ongoing knowledge of the project so that its members are well-placed to act should potential disputes arise. The recommendations proposed by the board should hopefully contribute towards the Parties continuing to work closely together, as opposed to the development of a dispute and the consequent drain on resources that brings.

Clause W3.1(7) states that the Dispute Avoidance Board members, along with their employees and agents, are not liable to the Parties for any action or failure to act in resolving a potential dispute. This is unless the action or failure to act was in bad faith, in which case they may open themselves up to some sort of liability. This matches the statement on the liability of the *Adjudicator* under the adjudication procedures in Options W1 and W2 and of dispute resolvers generally.

The Parties, conversely, must do the following.

- Act to appoint a third Dispute Avoidance Board member, where the Dispute Avoidance Board members are three in number (Clause W3.1).
- Decide the frequency of Site visits by the Dispute Avoidance Board members (Clause W3.1(5)).
- Propose the agenda for the Site visit (Clause W3.1(6)).

- Refer to the Dispute Avoidance Board a potential dispute that arises under or in connection with the contract (Clause W3.2(2)).
- Comply with the timescales for notifying and referring potential disputes to the Dispute Avoidance Board (Clause W3.2(3)).
- Make available copies of the contract, progress reports and other material they consider to be relevant to the Dispute Avoidance Board (Clause W3.2(4)).
- Attempt to resolve a potential dispute by the end of the Site visit (W3.2(5)).

So how can we read all of this together to see how the drafters help the Parties to avoid disputes in NEC4 contracts? We think there are several benefits in using the Dispute Avoidance Board.

- The focus of this process is dispute avoidance.
- Having one or three Dispute Avoidance Board members should help in any Dispute Avoidance Board voting that needs to take place during the process.
- Regular visits to the Site are known about and planned at tender stage and the frequency can be adjusted after the contract is awarded.
- The regular visits allow for familiarity with the contract and people, which should make the process that bit more efficient.
- The Parties are encouraged to resolve each potential dispute before the end of each Site visit, otherwise they will receive a recommendation for resolving it from the Dispute Avoidance Board. This threat should focus the minds of the Parties, as potential disputes cannot be bounced from Site visit to Site visit. This is a very attractive feature of Option W3, but it might involve a lot of effort on or before the Site visit by the Parties in doing all they can to resolve the matter or else be as helpful, clear and honest with the Dispute Avoidance Board on the day. Time should be allocated to pre-visit discussions where this can be shown to be helpful.
- In fairness though, if the Parties have exhausted all reasonable efforts to resolve each potential dispute and have failed, then the Dispute Avoidance Board really earns its money by recommending how to resolve it. There is no detail on what this means, but it is likely to be focused on the best interests of the disputing Parties and hopefully not of their advisers.

The recommendation for resolving a potential dispute is quite interesting. The use of the (undefined) term 'potential dispute' leaves something to interpretation. Parties to disputes are used to arguing over whether one has crystallised or not, using the guidance from the English case of Amec Civil Engineering.[vii] But the subjective nature of a potential dispute is probably helpful, allowing a broad range of commercial issues to be considered by the board if requested by a Party. We imagine that it would be the simplest of recommendations for one Party to stop pursuing the potential dispute as there is simply no merit in it, or else to use negotiation, mediation or conciliation processes to reach a resolution. It could also be, for example, that Party A pays Party B a sum of money; this should be put into effect using Clause 12.3 of the ECC. We do not think it desirable for the Dispute Avoidance Board members to recommend going straight to the *tribunal*, unless perhaps the potential dispute touches on something of public interest and the *tribunal* is litigation. The contract does not state that the recommendation is binding in any way. It also, interestingly, does not state that the Dispute Avoidance Board recommendation can or cannot be used in the *tribunal* later.

Once the Dispute Avoidance Board makes a recommendation, a Party has just four weeks to give a notification of dissatisfaction if it intends to refer the matter that it disputes to the *tribunal*. This does not mean that the Party must go to the *tribunal*, or even actually refer the dispute to the *tribunal*, just to give notice of an intention to refer it to the *tribunal*. We often see notices of dissatisfaction in respect of adjudicators' decisions served routinely, to keep open the possibility of a reference to the *tribunal* in the future by avoiding the four weeks' time bar. There is every chance that such tactics may be deployed here with Dispute Avoidance Boards. There is little that the Parties can do to guard against such conduct.

With the emphasis clearly on such terms as settle and resolve, Option W3 provides a good basis for nipping potential disputes in the bud as early as possible. It may be, in fact, that justice does not prevail in certain instances, whatever is meant by justice, but the Dispute Avoidance Board is about assisting the Parties to manage the contract and resolve disputes in real time.

NOTES

i   International Federation of Consulting Engineers.
ii Option W3.1, 'The Dispute Avoidance Board'.
iii HMG (Her Majesty's Government) (1996) Housing Grants, Construction and Regeneration Act 1996. The Stationery Office, London, UK. Amended by HMG (2009) Local Democracy, Economic Development and Construction Act 2009. The Stationery Office, London, UK.
iv Chartered Institute of Arbitrators (2019) Dispute Boards. https://www.ciarb.org/das/dispute-appointment-service/dispute-boards/ (accessed 22/01/2019).
v   Institution of Civil Engineers (2019) Dispute Resolution. https://www.ice.org.uk/knowledge-and-resources/professional-practice/ice-dispute-services/dispute-resolution (accessed 22/01/2019).
vi Royal Institution of Chartered Surveyors (2019) Dispute Boards. https://www.rics.org/uk/footer/dispute-resolution-service/conflict-avoidance-consultancy/dispute-boards/ (accessed 22/01/2019).
vii *Amec Civil Engineering Ltd* v *Secretary of State for Transport* [2005] EWCA Civ 291.

Gerrard and Waterhouse
ISBN 978-0-7277-6404-1
https://doi.org/10.1680/necrad.64041.139
ICE Publishing: All rights reserved

Chapter 14
The NEC4 Dispute Resolution Service Contract (DRSC)

The drafters of the DRSC say 'This contract should be used for the appointment of an adjudicator or Dispute Avoidance Board member to resolve disputes under an NEC4 contract.'[i] The origins of the DRSC lie in the NEC Adjudicator's Contract found in earlier suites of NEC contracts, but this has been renamed and published as the DRSC for the first time in NEC4 as the contract is intended to be used by more than just adjudicators. The NEC approach to engaging an *Adjudicator* or board member differs from most other contracts. NEC separates (i) the process to be followed and (ii) the appointment of the person(s) to do it. The DRSC is used for the second of these two tasks. Options W1, W2 and W3 set out the procedure to be followed by the various participants in the dispute resolution process; the DRSC sets out the obligations and entitlements of the neutral person.

98. Avoiding disputes using the DRSC

There really is nothing in the DRSC that is designed to directly assist the Parties in, say, an ECC contract in avoiding a dispute, or even in trying to reach some sort of settlement in the meantime. The DRSC simply provides clauses for the Parties to have their dispute decided by a *Dispute Resolver*. That said, of course, the very threat of commencing the formalisation of dispute resolution proceedings is sometimes enough to encourage those in dispute to have another go at settling.

99. Resolving disputes using the DRSC

The *Dispute Resolver* may be acting as an adjudicator in one of the NEC4 contracts or a Dispute Avoidance Board member in the ECC. The Dispute Avoidance Board is only provided in the ECC, this being probably the contract that gives the best fit with this form of dispute resolution. There may be one or three Dispute Avoidance Board members, the odd number decided by the drafters to make sure there is no stalemate in decision making.

The *Dispute Resolver* is required to act impartially;[ii] this is a specific requirement not given to any other role in any other NEC4 contract. This is usually a requirement of the law in those jurisdictions that have legislation on specific construction dispute resolution. We think this is a quite appropriate requirement for such a role, though the contract does not deal with events where this does not happen. It would be a sad day when a person acting in this role chose to act partially, but all things are possible, of course. While allegations of apparent bias are numerous in dispute resolution circles, we cannot recall a situation where actual bias has been proved. The test in English law was established in *Porter* v *Magill*[iii] and is 'whether those circumstances would lead a fair-minded and informed observer to conclude that there was a real possibility that the tribunal was biased'. To help achieve the aim of Clause 1.2, the *Dispute Resolver* is obliged to notify the Parties as soon as a matter arises that might present a conflict of interest.[iv] One would expect that the process of deciding the *Dispute Resolver* would, in the first place, have involved making sure that the proposed person can properly act in this role. Where appointments are made by a nominating body, the proposed person is usually required to sign a declaration that no conflict is known about at the time. Of course, a conflict may appear during the work in question so the obligation to declare a conflict is ongoing. Competence, neutrality, impartiality, experience, ability, reasonable fees, availability and good communication skills are the sorts of qualities that the Parties might wish to consider in advance of selecting the *Dispute Resolver*. As we have described, there could, of course, be more than a single person acting in this role. Even choosing your *Dispute Resolver* is not the most straightforward task! Use of a reputable nominating body will simplify the process for the Parties as the person proposed will have been suitably assessed before as being capable of being nominated.

The *Disputer Resolver's fee* is stated in the Contract Data. This could be per day, per hour, per visit to the Site (if required) and so on, but in most situations an hourly fee is usually proposed. Where dispute boards have been used under other forms of contract, members have typically been paid a 'retainer' for their ongoing routine work plus an hourly rate when additional obligations (such as dealing with a dispute) arise. The expenses that may be recovered are stated in Clause 1.5, though we would imagine that trying to itemise very small telephone, internet and data charge might not be worth the effort of the *Dispute Resolver* and that these would probably be waived. Interestingly, there are no accounts and records obligations or inspection rights, such as are found in other NEC4 contracts. Presumably, we must all implicitly trust the *Dispute Resolver*, although the general legal duty to demonstrate one's case still applies. The courts in England have heard occasional cases concerning the non-payment of *Adjudicator's* fees and sometimes these have referred to what was asserted to be unreasonably large fees.[v] None of those cases has succeeded.

Clause 3.9 of the DRSC requires that when the *Dispute Resolver* is a Dispute Avoidance Board member, the Parties pay the amount due in equal shares; Clause 4.1 states that an invoice is submitted to each Party for its share.

Clause 2.11 of the DRSC requires that when the *Dispute Resolver* is an adjudicator, the *Parties* pay 'the amount due in equal shares'. There is an inconsistency between this clause of the DRSC and Clause W2.3(8) of the ECC, which states 'The *Adjudicator* may in the decision allocate the *Adjudicator's* fees and expenses between the Parties.' The ability of the *Adjudicator* to allocate fees between the *Parties* is a requirement of the construction contract legislation in the UK. Clause 1.8 of the DRSC (conflict) would therefore seem irrelevant here.

It could be that the *Dispute Resolver* bargained for an advanced payment to be made, although several adjudicator nominating bodies prohibit their panel members from demanding one. The details of this are found in Clause 3.7, where the *Dispute Resolver* is a Dispute Avoidance Board. The repayment is included in the amount due assessed after the Site visit. Where the *Dispute Resolver* is to act as an adjudicator, Clause 2.9 provides that, whenever a dispute is referred to the *Dispute Resolver*, the Party referring the dispute pays the *Dispute Resolver* the amount stated. This is then reconciled in the amount due assessed after the decision on the dispute has been notified. Trying to get paid by a Party after a decision is made, where that Party is disappointed with the decision, sometimes proves difficult for an adjudicator, hence the advanced payment might appear quite attractive, certainly from the *Dispute Resolver's* perspective. The reality of adjudication is that in many cases the Parties merely ignore advanced payment provisions, knowing that the *Adjudicator* must comply with a timetable regardless of the *Parties'* acts or omissions.

There are no express provisions for what happens if either of the Parties disagrees with an invoice. Where payment is late, interest is paid on the late payment.[vi] Where one of the Parties fails to pay the *Dispute Resolver*, the other Party pays the amount due with interest. Clause 4.7 then provides that the defaulting Party repays the other Party the amount paid to the *Dispute Resolver* (which would include interest) together with some further interest. Many adjudicators face difficulties in collecting payments and some, as we mentioned earlier, must resort to litigation to secure payment. The reasons proffered for non-payment usually relate to the appointment and jurisdiction of the *Adjudicator*, rather than any suggestion that the decision was wrong, which is rarely a ground for challenge.

The *Dispute Resolver* and the Parties must observe the communications rules stated in Clauses 1.9 and 1.10 and must not do any Corrupt Acts, though the latter is not further mentioned in terms of remedies in any other clause. That said, the Parties may, by agreement, terminate the appointment of the *Dispute Resolver* for any reason. All they must do is have a reason, which a Corrupt Act certainly would be, and then notify the *Dispute Resolver* of the termination.[vii]

The *Dispute Resolver* also has the right to terminate the appointment for one of four reasons stated in Clause 5.2. 'Any reason' is not listed but 'unable to fulfil the role of *Dispute Resolver*' is. This allows the *Dispute Resolver* to resign in any number of difficult scenarios, usually caused by the Parties' own conduct, for example refusing to agree to reasonable requests for additional time after the receipt of voluminous submissions. More mundane reasons, such as holidays and other professional obligations, should have been considered by the *Dispute Resolver* before accepting the appointment.

We have covered the *Dispute Resolver's fee*, a few of the rules, getting paid and termination of the appointment, so what exactly does the *Dispute Resolver* do in terms of resolving any dispute?

100. Resolving disputes – adjudication

The DRSC specifies that Clause 2 only applies if the Contract Data states that the *Dispute Resolver* acts as an adjudicator.[viii] Clause 2.2 prohibits a repeat dispute, one which is the same or substantially the same as one previously referred to the *Dispute Resolver* or its predecessor. This isn't always initially apparent, particularly as the *Adjudicator* is new to the contract. Sometimes one of the Parties tries to avoid the constraint of Clause 2.2 by rephrasing the dispute. The other Party is likely to spot this and explain to the *Adjudicator*. Having investigated the complaint, the *Adjudicator* should resign if the dispute is a rerun of an earlier, decided, dispute.

The *Dispute Resolver* is obliged to decide on a dispute referred under the *contract between the Parties*;[ix] where and what that is will be stated in the Contract Data and will need to be made available to the *Dispute Resolver*. Clause 2.3 goes on to require the *Dispute Resolver* to decide and notify the Parties of the decision, again in accordance with the *contract between the Parties*.

The detail of what the Parties do, when they submit their statements of case, give evidence, attend any meetings and the like, is all contained in the *contract between the Parties*. The Parties could therefore be using ECC Option A incorporating Option W1; in which case all the timescales and detail that the Parties and the *Dispute Resolver* need to follow are stated in Option W1. All the DRSC is basically saying is that the *Dispute Resolver* must act as adjudicator in the contract into which the Parties have entered. There is, therefore, the potential for a conflict between the DRSC and the *contract between the Parties*, as we mentioned earlier, but Clause 1.8 gives precedence to the DRSC.

It could be that the *Dispute Resolver* needs help from others that the *Dispute Resolver* considers necessary in reaching a decision. This is rare but does occasionally happen. There is flexibility, designed to allow the procedure to work as smoothly as possible, but the *Dispute Resolver* must first notify the Parties.[x] The advice obtained from a third party should be shared with the Parties, who should be allowed to comment on it. Any failure to share such information, where the *Adjudicator* relies on that advice, would represent a breach of natural justice.[xi]

Clause 2.5 details the level of co-operation required of the Parties, Clause 2.6 deals with confidentiality and Clause 2.7 details how long the *Dispute Resolver* must keep documents that are provided by the Parties. The term 'documents' presumably includes electronic files as well as hard copies. Most adjudications, given the restricted time periods, rely on cloud computing services, download sites and emails. You would think the *Dispute Resolver* would prefer cloud-based documents rather than keeping masses of paper documents somewhere in the office for lengthy periods of time.

So, the DRSC simply sets up an appointment between the Parties and the *Dispute Resolver* to allow the *Dispute Resolver* to act as *Adjudicator* in the *contract between the Parties*, and to get paid for doing just that, with a few other simple rules thrown in. The real meat in the adjudication dispute resolution procedure is found in the *contract between the Parties*, not in the DRSC.

101. Resolving disputes – Dispute Avoidance Board

The DRSC specifies that Clause 3 only applies if the Contract Data states that the *Dispute Resolver* acts as a Dispute Avoidance Board member.[xii] Clause 3.3 states that the *Dispute Resolver* undertakes the duties of a Dispute Avoidance Board member in accordance with the *contract between the Parties*. This clause goes on to require the Parties to co-operate with the *Dispute Resolver*. Clause 3.4 requires the *Dispute Resolver* to collaborate with other members of the Dispute Avoidance Board; this is quite an interesting requirement. Whatever this might mean, it must be read and applied against the impartiality clause, Clause 1.2.

Clause 3.5 deals with confidentiality and Clause 3.6 details how long the *Dispute Resolver* must keep documents that are provided by the Parties. Again, you would think the *Dispute Resolver* would prefer cloud-based documents rather than keeping masses of paper documents somewhere in the office for lengthy periods of time.

So, the DRSC simply sets up an appointment between the Parties and the *Dispute Resolver* to allow the *Dispute Resolver* to act as a Dispute Avoidance Board member in the *contract between the Parties*, and to get paid for doing just that, with a few other simple rules thrown in. Again, the real meat in the Dispute Avoidance Board dispute resolution procedure is found in the *contract between the Parties*, not in the DRSC.

102. Summary

All the DRSC is doing is providing a contract between the *Dispute Resolver* and the Parties. When the *Dispute Resolver* is acting as *Adjudicator* or Dispute Avoidance Board member, it will need to have sight of the *contract between the Parties*, which details the precise duties of the *Dispute Resolver*. The DRSC is an important document in the dispute process and is likely to be closely scrutinised in any enforcement proceedings. The nature of the *Dispute Resolver's* role in law differs materially from the roles of other people appointed to provide services in a construction project, hence the need for a stand-alone contract, instead of, say, one of the professional service contracts. There really is nothing explicit in the DRSC that can be viewed as helping the Parties avoid disputes; of course, the threat of using the *Dispute Resolver* may sometimes be enough for the Parties to come to their senses. Should a dispute arise and be formalised, the DRSC merely gives clear rules on the appointment of the *Dispute Resolver*.

NOTES
i   DRSC, p. ii.
ii DRSC, Clause 1.2.
iii *Porter* v *Magill* [2001] UKHL 67; [2003] AC 357.
iv DRSC, Clause 1.3.
v   Most recently, at the time of writing, *The Vinden Partnership Ltd* v *Orca LGS Solutions Ltd & Or* [2017] EWHC B24 (TCC).
vi DRSC, Clause 4.6.
vii DRSC, Clause 5.1.
viii DRSC, Clause 2.1.
ix DRSC, Clause 2.3.
x   DRSC, Clause 2.4.
xi *Balfour Beatty Construction Ltd* v *The Mayor and Burgesses of the London Borough of Lambeth* [2002] EWHC 597 (TCC).
xii DRSC, Clause 3.1.

Chapter 15
Secondary Options

103. Avoiding disputes using secondary Options

Just how can disputes be avoided using secondary Options? Well, most of the ECC secondary Options involve some sort of risk between the Parties. Option X2, for example, is about who takes the risk of changes in the law after the Contract Date. Option X7, conversely, is about the financial risk that the *Contractor* takes if it achieves Completion after the Completion Date, i.e. the *Contractor* is late.

So, it is fair to say that when choosing the secondary Options, a more sensible use of the Options should lead to a more balanced contract and perhaps minimise the chance of disputes arising. We cannot emphasise enough that this type of consideration at the start of the contractual process can provide so much value to the Parties later in terms of dispute avoidance. Most disputes are commercial; they are just about money. So, if the *Client* places too much risk on the *Contractor* and that risk materialises, there is every chance that the *Contractor* will lose money and try to find reasons to blame and recover such monies from the *Client*. Surely this is a quite normal occurrence in business transactions? The *Client* needs to consider this quite carefully in assembling reasonable terms of contract, including the use or not of secondary Options. Otherwise, be careful what you wish for, as the choices you make at tender stage will often drive the behaviour of others. The *Client* should therefore think very carefully about all of this. The tragedy is that so many procurement processes, frequently regulated by statute, do not permit dialogue between prospective partners; surely that is unfortunate?

Looking at the ECC secondary Options in turn, we highlight those that we think offer opportunities to avoid disputes.

104. Option X1 – price adjustment for inflation (used only with Options A, B, C and D)

If Option X1 is incorporated in the contract, the risk (to some degree) of inflation is taken by the *Client*. We say 'to some degree', as it depends how closely the proportions used to calculate the Price Adjustment Factor and their respective indexes, stated in the Contract Data, relate to true inflation that occurs during the contract. This is particularly important when using Option A or B from the *Contractor's* perspective, as it is taking all of the financial risk of the tendered Total of the Prices adjusted by Option X1. In Option C or D, this risk is effectively shared between the Parties, as the target (the Prices) will be adjusted by Option X1; if this adjustment proves to be insufficient, the Parties will probably end up having less gain or more pain because of this.

Inflation is, of course, an economy-specific issue, so the use of the main Options together with Option X1 becomes a very important issue when preparing the tender documents. In countries such as Zimbabwe (with extraordinarily high inflationary rates in recent times), clients may have been encouraged to use Option E or F on any contracts, inflation probably being the most important financial threat to the success of these.

The right choice of main Option and the right choice of using Option X1 with the right proportions and indices will have a significant impact on getting a tick in the box for establishing appropriate Prices and the means to adjust these during the contract, helping the Parties avoid disputes on this.

105. Option X2 – changes in the law

Where Option X2 is included in the contract, knowledge of this will be of benefit to the *Contractor*, who does not have to guess at possible forthcoming changes of legislation and the impact that has on the project. The same, though, cannot be said of the *Client*, who takes that risk. The saying is that the risk is best placed with the person best able to manage it, which surely is the *Client*, who is doing the buying. Will this assist in helping the Parties to avoid disputes? Who knows really? It completely depends on the changes in legislation. Considering Brexit, which, at the time of writing, has troubled several organisations, do tenderers offer higher Prices or lower Prices if Option X2 is not included on those contracts being bid? Whoever knows the answer to this surely has a bright career in economics ahead of them!

106. Option X3 – multiple currencies (used only with Options A and B)

In its quite limited use, as we are aware, the main benefit that might reduce the likelihood of a dispute is in fixing the *exchange rate*. In certain countries, this will be a huge concern to tenderers so its inclusion with an appropriate *exchange rate* might help.

107. Option X5 – sectional Completion

In terms of helping the Parties to avoid a dispute, the inclusion of Option X5 makes it clear that the Completion of a certain part of the *works* in advance of the whole of the *works* is important to the *Client* and that the *Contractor* will need to plan and deliver the *works* carefully. Corresponding delay damages will often be added for failure to achieve this, in Option X7.

108. Option X6 – bonus for early Completion

Between us, we have occasionally seen contracts incorporating Option X6. The most common use is in commercial construction, for example retail premises, where the early Completion of a project will lead to additional revenue for the *Client* and, therefore, the bonus is paid for from additional revenue and there is no net cost to the *Client*. On the face of it, this is surely quite strange but, when you look at how we often set contracts up, the focus is more often about dealing with what happens when things go wrong than about rewarding a *Contractor* for improved performance. The performance being measured here is, of course, time and achieving Completion before the Completion Date. This can be used for any *section* or the whole of the *works*.

This Option would probably only be used where time was of the essence to the *Client*. Basically, the earlier they can put their shiny new asset into operational use, the quicker they will attain returns on their investment.

If the Completion Date is changed, perhaps because of a compensation event, the bonus is still preserved. This always allows the *Contractor* to seek to try to Provide the Works in the most time- and cost-effective way. The decision to invest resources or other measures to achieve the bonus will, of course, depend on the amount of the bonus. The *Client*, looking at very substantial returns once the asset is taken over, can afford to be more generous in setting the bonus in the first place.

The focus of the *Contractor* should be on achieving Completion as soon as it is reasonably able to do so. Where time is important to the *Client* and Option X6 is incorporated, surely there is a clear alignment of objectives here, so the Parties should be incentivised to work together to achieve each other's respective goals, thus reducing the likelihood of disputes?

109. Option X7 – delay damages

Option X7 is the reverse of Option X6. Where the *Contractor* achieves Completion after the Completion Date of any *section* or the whole of the *works*, it will pay delay damages until the earlier of Completion and the date on which the *Client* takes over the *works*.[i]

How might this help the Parties to avoid disputes? Well, it probably does not on balance, other than that the amount per day is known in advance, and certainly at tender stage, so the financial risk of this can and should be priced by the *Contractor*. We are (fairly) sure no sensible contractor would bid for a contract it is sure it could not provide by the date required; that would be unwise. But reasons for changing the Completion Date or not, generally through compensation events, form a very likely source of disagreement on contracts. The *Contractor* contends that the compensation event causes planned Completion to be delayed by two weeks and demonstrates this in its quotation.[ii] The *Project Manager* disagrees, finally making an assessment under ECC Clause 64 and assesses this as a one-week delay. The dispute, therefore, will often be about the time- and cost-effects of a compensation event, not the amount per day of delay damages stated in the Contract Data.

If Option X7 is not incorporated into a contract, the *Contractor*, of course, still does the work so that Completion is on or before the Completion Date, as provided in ECC Clause 30.1, but achieving Completion late will be a breach of contract. On such a contract, there would be uncertainty on how much this breach might amount to and whether the *Client* might pursue this.

The use of Option X7 might avoid a dispute over the amount per day of damages, but if this amount is higher, it should lead to increased Prices at tender stage and probably to a more defensive behaviour by the *Contractor* during the contract, who looks for reasons to get the Completion Date pushed out to something that can be achieved. Despite modern case law, challenges to the quantum of liquidated damages clauses still reach the courts.

110. Option X10 – information modelling

Will a clause on information modelling lead to a reduced likelihood of disputes arising? We think that information modelling has great potential to do just that. It is not the Option X10 drafting but instead the modelling process itself that should bring about significant improvements in outcomes. If all that information modelling achieved in practice were the removal of clashes in design, a rich source of an avoidable problem in the way that so many jobs are currently designed, this outcome alone would surely pay for information modelling. Note that the contract uses the term 'information modelling' and not 'BIM', as information modelling is applicable to other than just buildings.

The drafting covers collaborative working requirements, mandates further early warnings to be notified and always requires an updated Information Execution Plan to be in place; these positive requirements should help the successful implementation of information modelling.

A brave client may try one ECC contract using information modelling and a similar project not using it just to compare outcomes, but then we have been unwisely taking a two-dimensional disparate approach to design on a capital cost basis for just far too long now. We think that information modelling has the potential to significantly improve outcomes in our industry, which will surely lead to reduced disputes.

Conversely, information modelling introduces new risks about the use and security of data and these may lead to dispute. It also allows Subcontractors and external designers access to the data which, in turn, may lead to liability straying further down the supply chain.

Information modelling is a recent development in the industry and has been accompanied by a corresponding new problem in disputes. Once dispute proceedings are commenced, the Party controlling the data will block access to the common data for the other Party, thus making its advocacy in dispute resolution much more difficult. It is to be hoped that adjudicators use their authority to direct the restoration of access in these situations.

111. Option X12 – multiparty collaboration (not used with Option X20)

Option X12 cannot be used with Option X20 as Key Performance Indicators are included in both so there would be the potential for uncertainty to exist. If you do not want to use the multiparty clause but do want Key Performance Indicators, just use Option X20. If you do wish to promote multiparty collaboration, and on many projects surely this makes absolute sense, but you wish to have a mix of biparty and multiparty type Key Performance Indicators, this can easily be achieved using Option X12. Key Performance Indicators can be set solely on a target for the *Contractor*, where the target will be achieved according to the performance of several organisations. The Key Performance Indicators will need to be set up quite smartly at tender stage though. Because of the way that Option X12 (and Option X20) is drafted, there is nothing to stop the *Client* inserting Option X12 but having no Key Performance Indicators at tender stage, as these can be added by the *Promoter* after the contract is signed.[iii]

Will multiparty collaboration help to avoid disputes? Again, just like information modelling, you must believe that this will have tremendous potential to join up the dots in our very fragmented industry. Provided that all parties act ethically, having the earliest threats and opportunities made known to those that might be affected by them gives those parties the best chance of solving the threats and exploiting the opportunities. Why do we not do this as a matter of course on most projects? Who knows, but trust and money are likely to be the most significant barriers, as the likes of Latham[iv] and Tang[v] both alluded to some time ago.

112. Option X14 – advanced payment to the *Contractor*

In the authors' experience, most disputes are about money, the lack of it and the timing of it. When adjudication was being touted for introduction to the statute books in the UK, we recall statements such as that adjudication 'may be an element of rough justice' but is all about 'getting the right money in the right pocket sooner rather than later'. With so much emphasis (rightly) on money, what better way to start a project than to have an injection of finance upfront to avoid the *Contractor* having to finance the project? How sensible is that? The *Contractor* gets a better credit rating, they can reduce the Prices, as the financing element does not have to be included, and the *Client* pays a lesser amount, provided they can get the money cheaper than the *Contractor* can borrow it. But then we are building an asset for the *Client* so why should the *Contractor* have such a financing burden placed on it? Surely this is completely upside-down thinking? The sensible *Client* might well ask for alternative tenders, one with and one without Option X14, perhaps under Option A, and just test what positive effect this Option has on the Prices. If it does not drive the Prices down, do not use the Option, but if it does, then surely it makes absolute sense to use this as the *Client* will be protected by a bond for the advanced payment. The requirements of providing bonds do sometime mean that the cash benefit of advanced payments can be removed by a corresponding demand by the bond issuer for collateral from the *Contractor*.

So, will Option X14 help the Parties avoid disputes? We are quite sure that it will go a long way to achieving this, remembering always that 'Cash is king'[vi] and 'Cash flow is the life blood of the industry'.[vii]

113. Option X15 – the *Contractor's* design

Option X15 is now quite a neat clause that brings together a common set of requirements that the Parties often need when using design and build or just have elements of the *works* to be designed by the *Contractor*. The level of skill and care expected is stated; the use of material provided by the *Contractor* for other work, retention and return of documents and professional indemnity insurance are common issues to consider in *Contractor* designs.

Will this clause help the Parties to avoid disputes? Well, the requirements and expectations are quite clear; it really depends on the *Contractor* carrying this off as it basically promised it would do. There is no process in Option X15, just a set of rules, but if these are combined with the early warning and ambiguity and inconsistency clauses,[viii] together with a clear and well written Scope, the Parties have the basis of avoiding nasty surprises and potentially, therefore, disputes in this area. Disputes involving parties who dispute fit-for-purpose design obligations, or not, as the case may be, are not rare and the combination of NEC procedures may assist here. The contract, if poorly drafted, may set irreconcilable obligations on the *Contractor*. One such case,[ix] albeit not using an NEC contract, reached the UK Supreme Court in 2017, a rare visit to that court for the construction industry; that case demonstrating the difficulties in these situations.

Where used properly, this clause will help to avoid disputes, as it will make clearer what is expected of the *Contractor's* designers.

114. Option X17 – low performance damages

This clause compensates the *Client* in a contract where there is an output specification and the specified output or performance is not achieved. Just like Option X7 delay damages, the amount of damages should poor performance occur is known at tender stage. The likelihood can be considered by the *Contractor* and, knowing what the severity might be, sufficient risk monies, design solutions or construction techniques can be decided on.

This Option probably offers the Parties a quite limited opportunity to avoid disputes. This Option tends only to be used on process type plants, can be avoided by good design and construction and, even if that is not achieved, the breach starts life as a Defect, so the *Contractor* will generally have the chance to correct this and avoid such damages.

115. Option X18 – limitation of liability

This clause, again, will offer the Parties limited opportunity to avoid disputes. It will at least set out limitations of the *Contractor's* liability to the *Client*, which is rightly considered a serious threat to the long-term existence of a *Contractor's* business. Many people confuse limitation of liability with insurance. These are separate issues, both needing careful understanding, and are not particularly the subject of this book.

116. Option X20 – Key Performance Indicators (not used with Option X12)

We mentioned against Option X12 the positive incentivisation that Key Performance Indicators can bring to a contract. Key Performance Indicators are about continuous improvement so it is important that the drafting provides for incentives, not disincentives. We see many Z clauses dealing with turning Key Performance Indicators into negative payments for failure to meet the targets stated. We consider this a completely improper use of a very useful improvement tool.

117. Option X21 – whole life cost

A new Option introduced in some of the NEC4 contracts, this offers the *Contractor* the chance to forward a bright idea to save the *Client* in the operational costs of an asset. It may very well involve further capital expenditure, but the expected payback comes during the lifetime of the asset. This should be of considerable interest to owner occupiers, particularly where the *Contractor* was appointed via competitive tendering and played no part in producing the Scope. All of that said, it really is up to the original designer to produce something that meets the *Client's* expectations in the first place. Again, we must comment that ethics are not always on show and the *Client* may get a design that reflects a fee secured at a stupid level on a competitive basis.

Hopefully, some good, fresh ideas may be able to be brought to the project early enough to be beneficial to both Parties, increasing trust levels and thus helping to reduce the likelihood of a dispute arising.

118. Option X22 – early *Contractor* involvement (used only with Options C and E)

If only most projects could be negotiated! What a mature way of agreeing for one party to spend a particularly large amount of money with another party. The demands and benefits of competitive tendering can be combined with a collaborative approach to project development with early *Contractor* involvement.

> I would like the following to be designed and built and I have this amount of money for the capital costs and it is important that it costs no more than this amount of money in operational costs.

> Well, we have done this type of project before and, in the location you are thinking of and at the very testing levels of performance required, you may well struggle to achieve all of your goals. But have you thought about this…?

> That is very interesting; we did consider that a while ago. Tell me more.

And on it goes, where eventually the Parties conclude a deal going forward in terms of time, cost and quality, or they do not, in which case the *Client* has the right to go back out to tender if they truly believe they can get the deal they want. Of course, the *Contractor* will be extremely keen to conclude a deal for Stage Two during Stage One, and why not? Honesty, integrity and performing to your professional best is all you can do, and all the *Client* can expect.

119. Option Y(UK)1 – project bank account

Money, or at least the lack of it, is probably the big-ticket cause of disputes in our industry. NEC4 contracts offer up several tools to improve the normally very predictable problems associated with payment. Various incentives can be used, such as Key Performance Indicators. A target cost arrangement can be used; advanced payment can be made; no retention is an option; and reduced certification and payment periods can also be decided by the *Client*. A project bank account gives the Parties a quite dramatic alternative to the usual slow flow of payments down the supply chain. Although there seems to be a limited number of banks that offer this as a service,[x] most participants in a construction project can be paid at worst in a few weeks after the assessment date rather than sometimes several months after this and with protection from insolvency higher up the chain. Surely, that is a quite incredible feat for an industry beset with money issues? How refreshing it is to see a tool created to get the right money in the right pockets sooner rather than later.

The very fact that the traditional chasing of money becomes less of an issue should lead to fewer disputes. We are not aware of any disputes arising regarding money on projects using a project bank account. Perhaps this is another area worthy of academic study.

120. Option Y(UK)2 – Housing Grants, Construction and Regeneration Act 1996

The main benefit of this clause from a UK perspective is that those adopting this clause where the Act is applicable to their project can safely use it in the knowledge that they are Act-compliant.[xi] We have heard of no legal commentators who have any issues with this clause; therefore, the Parties are at least spared the risk of a dispute arising from this.

121. Option Z – additional conditions of contract

The only thing to comment on here is that the *Client* may very well introduce this Option, commonly known as 'Z clauses', which could possibly bring in further clauses that might help to avoid disputes, but we generally find that these usually have the opposite effect, creating uncertainty and confusion, and can become a source of dispute in themselves. Not many commentators have good things to say about Z clauses; these should be drafted with care and both understood and priced accordingly by *Contractors*.

There are plenty of examples of ill-conceived and poorly drafted Z clause that have the potential to cause major disputes. Some further examples and commentary on these can be found in the *Z clauses Laid Bare* series found in *NEC Contracts, Official Group – Developers and Publishers of NEC3 & NEC4 Suites of Contracts*.[xii]

122. Resolving disputes using secondary Options

We have outlined those secondary Options that offer users the chance to avoid disputes to varying degrees, but few of them also provide for a process to resolve disputes. The exception to this is Option X12, on multiparty collaboration, and possibly any Z clauses. We have seen several Z clauses introduced to create a dispute escalation ladder; Wong sets out some thoughts on this in part 36 of his paper.[xiii] Quite popular in the UK, certainly with public sector bodies, several organisations set out a process to gradually escalate disputes from the quick and cheap negotiation, through mediation and adjudication to, finally, arbitration or litigation. We have even seen this where contracts are caught by the Housing Grants, Construction and Regeneration Act 1996, as amended, one of the clauses generally being drafted to say that rights under the Housing Grants, Construction and Regeneration Act 1996 are unaffected by this process. Which, of course, they are. The hope is that most disputes can be resolved as cheaply and quickly as possible, in a form where the Parties retain control of the outcome.

Option X12 does not specifically mention disputes or dispute resolution, but surely one of the features of using this clause is to benefit from an open, multiparty collaborative way of working and one would expect that the Core Group acts in an adult way to facilitate early resolution of potential disputes. We do not think that this needs to be stated as a function of the Core Group, it is so obvious.

123. Summary

There are several secondary Options that, when selected, offer the Parties the chance to avoid situations that might otherwise lead to disputes. This shows the importance of carefully drawing up a contract to align the objectives of the Parties. As we have said elsewhere in this book, the construction industry is set up to deliver projects for clients; it is not, and should not be, about pursuing remedy for disputes that arise. In England and Wales, the industry has its very own court (the Technology and Construction Court[xiv]) to deal with the volume and speciality of construction disputes.

NOTES
  i   ECC, Clause X7.1.
  ii ECC, Clause 63.5.
  iii See ECC Clause X12.4(2).
  iv Latham M (1994) *Constructing the Team*. HMSO, London, UK. See http://constructingexcellence.org.uk/wp-content/uploads/2014/10/Constructing-the-team-The-Latham-Report.pdf (accessed 22/01/2019).
  v   Construction Industry Review Committee (2001) *Construct for Excellence: Report of the Construction Industry Review Committee*. Development Bureau, Hong Kong. See https://www.devb.gov.hk/en/publications_and_press_releases/studies_and_reports/report_of_the_construction_industry_review_committee/index.html (accessed 22/01/2019).
  vi Kenton W (2018) Cash is king. In *Investopedia*. Dotdash, New York, NY, USA. https://www.investopedia.com/terms/c/cash-is-king.asp.
  vii So said Lord Denning in a series of Court of Appeal decisions, starting with *Dawnays* v *Minter* (1971) and finishing with *Modern Engineering* v *Gilbert Ash* (1973).
  viii ECC, Clauses 15 and 17, respectively.
  ix *MT Højgaard A/S* v *E.ON Climate & Renewables UK Robin Rigg East Limited and another* [2017] UKSC 59.
  x   Barclays Bank and the Royal Bank of Scotland currently offer project bank accounts.
  xi HMG (Her Majesty's Government) (1996) Housing Grants, Construction and Regeneration Act 1996. The Stationery Office, London, UK. Amended by HMG (2009) Local Democracy, Economic Development and Construction Act 2009. The Stationery Office, London, UK.
  xii Thawrani R (2019) NEC Contracts, Official Group - developers and publishers of NEC3 & NEC4 suites of Contracts. In LinkedIn. https://www.linkedin.com/groups/3378841 (accessed 22/01/2019).
  xiii Wong W (2016) Keynote address. *IBA Conference on 'Mediation v Arbitration: Best friends or Best enemies? A view from Asia', Hong Kong*. International Bar Association, London, UK. https://www.doj.gov.hk/eng/public/pdf/2016/lo20161201e1.pdf (accessed 22/01/2019).
  xiv Her Majesty's Courts & Tribunals Service (2019) Technology and Construction Court. https://www.gov.uk/courts-tribunals/technology-and-construction-court (accessed 22/01/2019).

Index

abbreviations, xi
acceleration, 36
Accepted Programme
    and compensation events, 72–73, 94–95
    definition of, 31
access date(s), 30, 72
Activity Schedule, 49–52, 61–62
adjudication
    communications in, 124, 130–131
    of compensation events, 99–100
    process of
        in Option W1, 47–48, 122–124
        in Option W2, 47–48, 129–131
        in Option W3, 47–48, 133–137
Adjudicator
    and adjudication process
        in Option W1, 47–48, 122–124
        in Option W2, 47–48, 129–131
    authority of, 123–124, 130
    decisions of, 124, 131
    Dispute Resolver acting as, 141–142
    fees of, 140
    impartiality of, 122, 128
    replacement for, 122, 128–129
    role of, 7
    selection/nomination of, 122, 128
Adjudicator nominating bodies, 122, 128, 140
advanced payments
    to *Contractor*, 88, 148–149
    of *Dispute Resolver*, 140
Alliance Contract (ALC)
    in general, 107, 108
    use of *Senior Representatives* in, 117

Amec v Secretary of State for Transport case, 17–18, 136
amount due
    assessment of
        in general, 40–41
        corrections to, 42–43, 48
        final, 41, 44–48
        intervals, 40–41
arbitration, advantages/disadvantages of, 111–115
Arcadis, 113
assessment
    of amount due
        in general, 40–41
        corrections to, 42–43, 48
        final, 41, 44–48
        intervals, 40–41
    of compensation events, 92–96
    of Defined Cost, 54–61
    of delays, caused by compensation events, 32, 94–95
authority
    of *Adjudicator*, 123–124, 130
    delegation of, 14
    of *Project Manager*, 13, 14
    of *Supervisor*, 14
avoidance of disputes
    in general, 1
    arbitration or litigation in, 114–115
    and Dispute Resolution Service Contract, 139
    in Option W1, 121
    in Option W2, 127
    by using secondary Options, 145–152
    see also Dispute Avoidance Board

Bills of Quantities, 52–54, 62, 85–87
bonuses, for early completion, 146–147
Building Information Modelling (BIM), 15–16

certificate, use of term, 13
change, disputes and, 5–6
clauses
    choice of, 2–3
    see also core clauses; main Options;
        secondary Options; *under specific*
        Options
Client
    and access to Site, 72
    breach of contract by, 83
    choice of clauses by, 2–3
    choice of contract by, 2
    and compensation events, 70, 72–74, 79, 81,
        82, 83
    and early warnings, 22
    and final assessment, 46–47
    and Key Dates, 37
    liabilities of, 6, 81
    and materials for tests, 82
    and *method of measurement*, 53
    and payment processes, 43–45
    and terminations, 104–105
Client's Scope, 71
collaboration, 148
collateral warranties. *see* Option X clauses, X8
communication systems, electronic, 10, 11–12
communications, 9–19
    in general, 9–11
    in adjudication, 124, 130–131
    changes in, 10
    in disputes, importance of, 16–18
    ECC's requirements for
        Clause 13, 10–15
        Clause 14.3, 15
    and electronic communication systems, 10,
        11–12
    vs notifications, 9–10
    physical transmission of paper, 12
    rejection of, 13
    responses to, 13, 14
    systems for, 10, 11–12
    timescales in, 9, 31
compensation events
    in general, 6, 69
    in adjudication/at *tribunal*, 99–100

    in contract, 70–71
    delays caused by, 32, 94–95
    and early warnings, 26–27
    and force majeure clause, 83–84
    overview of
        in core clauses, 69–85
        in main Options, 85–87
        in secondary Options, 87–88
    and physical conditions, 78–81
    processes for
        in general, 69–70, 89
        assessment, 92–96
        implementation, 98
        notifications in, 89–91
        quotations in, 91–92
        quotations in, responses to, 84, 97–98
    and programme, 32
    and take over, 82
    and weather, 79–81
Completion
    and compensation events, 83–84
    definition of, 31
    and delay damages, 147
    early, 146–147
    and final assessment, 46
    planned, 31, 32, 94
    sectional, 146
    and take over, 35–36, 82
Completion Date
    definition of, 31
    and delay damages, 147
    and take over, 35–36, 82
condition, use of term, 37
conditions of contract, 75
Construction Contracts (Northern Ireland)
    Order 1997, 107, 108
construction contracts, UK legislation on, 48–49
Contract Data, 30, 72
Contractor
    access to Site of, 34–35
    and Activity Schedule, 50–51
    advanced payments to, 88, 148–149
    and assessment of amount due, 40–41
    and communication, 13
    and compensation events
        assessment of, 96
        and force majeure, 83–84
        notifications in, 89–91
        and physical conditions, 78–81

quotations in, 91–92
quotations in, responses to, 84, 97–98
and Defects, correction of, 88
and Defined Cost, 59–60
and Disallowed Cost, 57–58
early involvement of, 150
and early warnings
    in general, 22–25
    failure to give, 25–28
and final assessment, 46–47
and Key Dates, 37–38
liabilities of, 6, 15, 149–150
and *method of measurement*, 53
notifications to, 82–83, 84
and payments
    in general, 43–45
    in advance, 148–149
and programme
    in general, 29–32
    keeping up to date of, 33–34
and search for Defects, 77–78
and terminations, 104–105
Contractor's design, 88, 149
Contractor's Scope, 71
contracts
    breach of, 83
    choice of, 2
    cost reimburseable, 62–63
    dispute boards in, 133
    management, 63–64
    target, 61–62
    see also Alliance Contract (ALC); Dispute Resolution Service Contract (DRSC); Engineering and Construction Contract (ECC); Framework Contracts; NEC4 contracts; Supply Short Contract (SSC)
core clauses
    Clause 10.1, 26
    Clause 11.2, 27, 31, 49–50, 52–54, 63–64
    Clause 13, 10–15
    Clause 13.4, 76, 77
    Clause 13.8, 76
    Clause 14.1, 76
    Clause 14.3, 15
    Clause 15, 21, 22–25
    Clause 20.1, 102
    Clause 20.4, 59
    Clause 25.1, 34, 73
    Clause 25.2, 73
    Clause 25.3, 37–38, 75
    Clause 26.4, 59
    Clause 30.2, 75
    Clause 31.1, 29–30
    Clause 31.2, 31–32, 73
    Clause 31.4, 50–51
    Clause 33.1, 35
    Clause 34.1, 73
    Clause 35.3, 82
    Clause 36.1, 77
    Clause 41.2, 82
    Clause 41.5, 78
    Clause 43.1, 77
    Clause 45.1, 71
    Clause 45.2, 71, 77
    Clause 50.1, 40–41, 75
    Clause 50.2, 59
    Clause 50.3, 41–42
    Clause 50.5, 29–30
    Clause 50.6, 42
    Clause 50.9, 66
    Clause 51.1, 43
    Clause 52.1, 60
    Clause 52.2, 59
    Clause 52.4, 59
    Clause 53.3, 47, 65–66
    Clause 53.4, 48
    Clause 55.1, 50
    Clause 56.1, 53
    Clause 60.1, 33, 36, 70–84, 90, 97
    Clause 60.2, 79
    Clause 60.3, 79
    Clause 60.4, 85–86
    Clause 60.5, 86
    Clause 60.6, 86
    Clause 60.7, 54, 86–87
    Clause 61, 90–91
    Clause 61.2, 124, 131
    Clause 61.3, 90
    Clause 61.5, 26, 75
    Clause 61.6, 83, 91–92
    Clause 62, 91–92
    Clause 63.1, 93
    Clause 63.17, 75
    Clause 63.5, 32, 94–95
    Clause 63.6, 83, 89
    Clause 63.7, 26
    Clause 63.8, 95
    Clause 64, 96

Clause 64.1, 75
Clause 65, 84, 96, 98
Clause 90.1, 101–102
Clause 90.2, 102
Clause 90.3, 102–103
Core Group, 87
cost reimburseable contract, 62–63
costs
    and early warnings, 22–23
    finalisation of, 47, 65–66
currencies, multiple, 146

decisions
    of *Adjudicator*, 124, 131
    of *Project Manager*, 75–76
    use of term, 75
Defects
    corrections of, 88
    search for, 77–78
Defects Certificate, 46
defects date, 46
Defined Cost
    assessment of, 54–61
    definition of, 93
    rules for, 44–45
delay damages, 147
delays
    in general, 33
    in advanced payments, 88
    assessment of, caused by compensation events, 32, 94–95
delegation, of authority, 14
disagreements, vs disputes, 17
Disallowed Cost, 27–28, 56–58, 64
Dispute Avoidance Board, 133–137
    in general, 107
    benefits of using, 136
    members of
        Dispute Resolver acting as, 142
        selection/nomination of, 133–134
        Site visits of, 135
        skill set of, 134
    recommendations of, 136–137
    role of, 7, 135–136
    and UK legislation, 133
Dispute Avoidance Board nominating bodies, 134
dispute boards, 133
Dispute Resolution Service Contract (DRSC), 107, 122, 128, 134, 139–143

Dispute Resolver
    acting as
        Adjudicator, 141–142
        Dispute Avoidance Board member, 142
    fees of, 140–141
    impartiality of, 139
    right to terminate appointment, 141
    role of, 139–140
disputes
    and Activity Schedule, 52
    avoidance of
        in general, 1
        arbitration or litigation in, 114–115
        and Dispute Resolution Service Contract, 139
        in Option W1, 121
        in Option W2, 127
        by using secondary Options, 145–152
        see also Dispute Avoidance Board
    change and, 5–6
    communications in, importance of, 16–18
    definitions of, 17–18
    vs disagreements, 17
    potential, 136
    procedures in NEC4 for, 107–110, 109t
    programmes' role in, 33
    resolving of
        in general, 7
        about final assessments, 47–48
        arbitration or litigation in, 111–114
        and Defined Cost, 61
        and Dispute Resolution Service Contract, 139–143
        and Option W1, 47–48, 121–125
        and Option W2, 47–48, 127–132
        and Option W3, 47–48, 133–137
        and terminations, 106
disruptions, 33
DRSC (Dispute Resolution Service Contract), 107, 122, 128, 134, 139–143

early involvement, of *Contractor*, 150
early warning meetings, 23–24
Early Warning Register, 22, 23–24
early warnings, 21–28
    in general, 21–23
    application of, 24–25
    and compensation events, 26–27
    failure to give/act on, 25–28

and Parties' relationships, 28
process of, 23–24
triggers for, 22–23
Engineering and Construction Contract (ECC)
in general, 1–8
BIM requirements in, 15–16
clauses in. *see* clauses
communications in. *see* communications
compensation events in. *see* compensation events
early warnings in. *see* early warnings
liabilities in, 6, 149–150
objectives of, 1–2
payment provisions in. *see* payments
people in, 3
programmes in. *see* programmes
quality provisions in, 4–5
relationships with third parties in, 3–4
resolving disputes provisions in, 7
secondary Options in. *see* secondary Options
structure of, 2–3
termination in. *see* termination
terminology in, 30–31
time management in, 4

Fee, 45
fee percentage, 45
fees, 140–141
float, 32
force majeure clause, 83–84
Framework Contracts, 107

Housing Grants, Construction and Regeneration Act 1996, 48–49, 88, 107, 108, 151

impartiality/neutrality
of *Adjudicator*, 122, 128
of *Dispute Resolver*, 139
of *Senior Representatives*, 118
implementation, of compensation events, 98–99
inflation, price adjustment for, 145–146
information modelling, 147–148
interest, charging of/payment of, 43

Key Dates
in general, 37–38
changes to, 15
definition of, 30
Key Performance Indicators, 150

legislation
changes to, 87, 146
in UK, 48–49
liabilities, 6, 15, 81, 149–150
litigation, advantages/disadvantages of, 111–115
low performance damages, 149

main Options
Option A clauses, 49–52
Option B clauses, 52–54, 85–87
Option C clauses, 54–62, 65–66
Option D clauses, 54–61, 62, 65–66, 85–87
Option E clauses, 54–61, 62–63, 65–66
Option F clauses, 63–64, 65–66
Option W1 clauses. *see* Option W1 clauses
Option W2 clauses. *see* Option W2 clauses
Option W3 clauses. *see* Option W3 clauses
see also secondary Options
management contract, 63–64
method of measurement, 53
multiparty collaboration, 4, 21–22, 148
multiple currencies, 146

NEC4 contracts
dispute procedures in, 107–110, 109t
see also under specific contracts
neutrality. *see* impartiality/neutrality
New Statesman, 113
nomination. *see* selection/nomination
notifications
vs communications, 9–10
and compensation events, 89–91
drafting of provisions for, 91
use of term, 13

objectives, of ECC, 1–2
opening statement, of ECC, 2
Option A clauses, 49–52
Option B clauses, 52–54, 85–87
Option C clauses, 54–62, 65–66
Option D clauses, 54–61, 62, 65–66, 85–87
Option E clauses, 54–61, 62–63, 65–66
Option F clauses, 63–64, 65–66
Option W1 clauses
in general, 47–48, 108
avoiding disputes in, 121
resolving disputes in, 121–125
use of *Senior Representatives* in, 117

157

Option W2 clauses
   in general, 47–48, 108
   avoiding disputes in, 127
   resolving disputes in, 127–132
   use of *Senior Representatives* in, 117
Option W3 clauses
   in general, 47–48
   Dispute Avoidance Board in, 133–137
Option X clauses
   X1, 96, 145–146
   X2, 87, 146
   X3, 146
   X5, 146
   X6, 146–147
   X7, 36, 147
   X8, 4
   X10, 15–16, 147–148
   X11, 102
   X12, 4, 21–22, 87, 148
   X14, 88, 148–149
   X15, 88, 149
   X17, 149
   X18, 149–150
   X20, 150
   X21, 150
   X22, 150
Option Y clauses
   Y(UK)1, 151
   Y(UK)2, 48–49, 88, 151
Option Z clauses
   in general, 3, 66, 151
   and compensation events, 70
   and programme, 29–30

Parties. see Client; Contractor; Project Manager
payments, 39–67
   in advance, 88, 140, 148–149
   processes for
     in general, 5, 39–40
     in core clauses, 40–48
     in main Option clauses, 49–66
     in secondary Option clauses, 48–49, 64, 66
   and terminations, 105–106
   see also Defined Cost; Disallowed Cost; Fee; Price for Work Done to Date
people, in ECC, 3
period for reply, 13
physical conditions, 78–81
planned Completion, 31, 32, 94

Porter v Magill case, 139
price adjustment, 145–146
Price for Work Done to Date
   in general, 41–42
   and Activity Schedule, 50
   and Bill of Quantities, 54
   definition of, 54–55, 63
Prices/*prices*, 63, 93–94
procedures
   for disputes, in NEC4 contracts, 107–110, 109t
   on termination, 104–105
programmes, 29–38
   in general, 29–30
   changes to, 34
   contents of, 31–32
   keeping up to date of, 33–34
   role in disputes, 33
   terminology of, 30–31
   see also Accepted Programme
project bank account, 151
Project Manager
   and Activity Schedule, 50–51
   and assessment of amount due, 40–41
   authority of, 13, 14
   belated replies by, 74
   and communication, 13–14
   and compensation events
     assessment of, 96
     and changes to Scope, 71
     and declining of acceptance, 76–77
     notifications in, 89–91
     and objects of value, 74–75
     quotations in, 91–92
     quotations in, responses to, 84, 97–98
     and revision of decisions, 75–76
     and stating of assumptions, 82–83
     and to stop/not start work, 73
     and take over, 82
     time restrictions in, 98–99
   and Defined Cost, 54–55, 59
   and Disallowed Cost, 57–58
   and early warnings
     in general, 22–25
     failure to act on, 25–28
   and final assessment, 46–47
   instructions of
     for changing Scope, 15, 71
     for dealing with objects of value, 74–75
     to stop/not start work, 73

and Key Dates, 15, 37–38
and *method of measurement*, 53
notifications of, to *Contractor*, 82–83, 84
obligations of, 75
and payment processes, 43–45
and programme, 34
and take over, 35–36
and terminations
  in general, 104–105
  role in, 103–104

quality, 4–5
quality plan, 32
quotations
  and compensation events
    in general, 91–92
    responses to, 84, 97–98

receipt, acknowledgement of, 10
rejection, of communications, 13
replacement, for *Adjudicator*, 122, 128–129
resolving of disputes
  in general, 7
  about final assessments, 47–48
  arbitration or litigation in, 111–114
  and Defined Cost, 61
  and Dispute Resolution Service Contract, 139–143
  and Option W1, 47–48, 121–125
  and Option W2, 47–48, 127–132
  and Option W3, 47–48, 133–137
responses
  to communications, 13, 14
  to quotations, 84, 97–98
risk allowance, 94–95
role
  of *Adjudicator*, 7
  of Dispute Avoidance Board, 7, 135–136
  of *Dispute Resolver*, 139–140
  of *Project Manager*, 103–104
  of *Senior Representatives*, 7, 118–119

sanctions, 25–26
Schedule of Cost Components (SCC), 60–61, 64–65, 93
Scope, changes to, 15, 71
secondary Options
  in general, 7
  Option X clauses. *see* Option X clauses
  Option Y clauses. *see* Option Y clauses
  Option Z clauses. *see* Option Z clauses
  and payments, 64
  see also main Options
sectional Completion, 146
selection/nomination
  of *Adjudicator*, 122, 128
  of members of Dispute Avoidance Board, 133–134
  of *Senior Representatives*, 118
Senior Representatives, 117–120
  in general, 47, 107
  in different contracts, 117
  neutrality of, 118
  role of, 7, 118–119
  selection/nomination of, 118
Short Schedule of Cost Components, 64–65, 93
  see also Schedule of Cost Components (SCC)
Site, access to, 34–35, 72
sole remedy clause, 83, 89
SSC (Supply Short Contract), 107
starting date, 23, 30
structure, of ECC, 2–3
Subcontractors, 24
Supervisor
  authority of, 14
  and communication, 14
  and compensation events, 75–76, 77–78, 90–91
  and final assessment, 46–47
  instructions of, 77–78
  tests or inspections by, 78
Supply Short Contract (SSC), 107
Sutcliffe v Thackrah case, 40
swimming pool case, 112

take over
  in general, 35–36
  and compensation events, 82
  and Completion, 35–36, 82
  definition of, 31
target contract, 61–62
termination, 101–106
  of contracts
    in general, 101
    definition of, 101–102
    in ECC, 6–7, 101
    for convenience, 102
    payments due on, 105–106
    procedures on, 104–105

Project Manager's role in, 103–104
   reasons for, 103
   starting of, 102–103
Termination Table, 102, 105
terminology, 30–31
time, terminology of, 30–31
time management, 4
time risk allowances, 32
timescales, in communications, 9, 31
Trant v Mott MacDonald case, 12
tribunal, 47, 99–100, 107

UK Ministry of Justice, 113
United Kingdom
   legislation on construction contracts in, 48–49
   methods of measurement in, 53

Waller, H., 113
weather, 79–81
whole life cost, 150
Wong, W., 151